NB

PASSIONS AND PREJUDICE

SPECIAL EDITION

Passions and PREJUDICE

THE SECRETS OF SPINDLETOP

Linda Light

PASSIONS AND PREJUDICE

Expanded, Special Edition
Copyright 1997 Linda Light

Published by Spindletop Productions, Inc.

ISBN 0-9645617-2-7

Without limiting the rights under copyright reserved above, no part of this publication may be reproduced, stored in or introduced into a retrieval system, or transmitted, in any form or by any means (electronic, mechanical, photocopying, recording or otherwise) without the written express permission of both the copyright owner and the above publisher of this book. All rights reserved.

The characters in this book are real historical figures, and Spindletop itself is a true historical event. Limited fictionalization, however, has been used at times for strictly entertainment purposes to capture and point up the meaning and drama of this story more clearly, as seen fit by the author, and as interpreted entirely by the author's imagination and personal read on the research done.

Every effort has been made to secure correct permissions for articles and to secure correct permissions to reprint photographs. If necessary, should there have been oversights, omissions, or corrections, these will be gladly added in future printings.

To my children Christa, Danielle, and Andrea
and
my Miles Frank,
with love always.

TABLE OF CONTENTS

Preface and Acknowledgments XI

Foreword XXV

Part I: Beginnings 1

Part II: The 13th of Spindletop 29

Part III: Blue Blood 97

Epilogue 231

Spindletop Now 237

Appendix I: A Dollar Bought a Lot
                    in 1930 253

Appendix II: Spindletop Hall and
                    Spindletop Farms at their Height ... 257

PASSIONS AND PREJUDICE

TABLE OF CONTENTS, continued

Appendix III: Reality Bytes from Spindletop
      Times: Select Oral Histories 277

    1.   Mr. and Mrs. Fred Wachs, Jr.,
    Lexington, Kentucky 279

    2.   Mr. George Stanhope Wiedemann,
    Lexington, Kentucky 309

    3.   Mr. and Mrs. Jack Grant,
    Outside of Texarkana, Texas 311

    4.   Mr. Marion J. Anderson,
    Hope, Indiana 341

    5.   Trainer Art Ledbetter,
    Henderson, Kentucky... 343

Appendix IV: Core Documents from
      Spindletop Times 345

About the Author 403

**A 64-Page Photo Section Is Inserted Between
Pages 254 and 255.**

PREFACE AND ACKNOWLEDGMENTS

Above all else, this book has been a search for the truth, and a painstaking fitting of puzzle pieces together to capture and crystallize the essence of the story of Spindletop and its lady--its marrow and meaning and drama, given the lapse of thirty-five years since its lady's death, given the sometimes conflicting recollections and differing perceptions of those still living from that time, and given the scarcity of records, many of which have been buried with time. It is also an attempt to give a woman ahead of her times her rightful place in history. I have not tried to gloss over history or its darker side, but to follow the historical lines of Spindletop and its lady's story, its signposts, and its "insides" as closely as I can, and as I understand them.

PASSIONS AND PREJUDICE

This book is a result of years of in-depth research and hundreds of live interviews with people in Texas and Kentucky, many of whom knew Pansy Yount personally or were contemporaries of the Spindletop times. All of these contributions come together here to form the opinions and findings in this book in an attempt to get to the soul of a saga that changed the lives of people everywhere. I wish to thank the fine people from Kentucky and Texas who shared their recollections with me and contributed to the making of this book on the human story behind Spindletop. Many of those people remain nameless as people who have come up to me after performances and given me their insights and comments, and who have asked thought-provoking questions. Others, with whom I have had the honor of spending more time, have their names indelibly printed in my memory for their generously given time and their sharing that has made this book possible.

I wish to especially tha... ...oving husband, who gave me unconditional support during the long process--almost ten years--of compiling materials for and writing this book, as well as during the performing of the one-woman play, *Pansy Yount of Spindletop*.

I wish to thank my "agent" and dear friend Clara Taylor Spencer Houlihan Wiedemann,

PREFACE AND ACKNOWLEDGMENTS

lovingly known as "Spence," and her creative genius of a husband and who is an artist within his own right: George Stanhope Wiedemann, of Lexington, Kentucky. Both have been with me giving support, encouragement, and artistic help since the inception of this work. Spence and "Hope" both hold the record for the number of times two people can sit through performances of *Pansy Yount of Spindletop*. They have been no less than two Kentucky angels on special assignment. Thank you for your friendship.

Heartfelt gratitude goes to my dear friend, Kathryn Manion Haider of Chicago, granddaughter of Pansy Yount, for the hours we spent going through family "stuff" and xeroxing ourselves to death. I cherish being "the Yount family historian" and appreciate the opportunity and honor of doing this work on Spindletop. Thank you, too, Kathryn for pictures from the family collection for this book. And thanks again for the warm welcome you gave me in Chicago!

And there is no way to ever express my appreciation to David E. Brown of Louisville. Simply and inadequately put, thank you!

My undying appreciation to Ann Detlefs Minch of Louisville for being the beacon of light at the end of my darkest tunnel.

XI

PASSIONS AND PREJUDICE

My warmest appreciation also goes to the following, and many others for their kind sharing of recollections and/or materials and help in making this work possible:

The late Governor A.B. "Happy" Chandler, one of Kentucky's most colorful governors;

Dr. Frank Peterson, former Comptroller, University of Kentucky, and one of those responsible for acquiring Spindletop for the University of Kentucky;

Dr. Frank G. Dickey, former President of the University of Kentucky, Lexington;

Christie Marino, Curator of the Spindletop/Gladys City Museum of Lamar University in Beaumont, Texas;

A real "Texas Tornado" named Betty Lou Allison and her fine husband Jim--also known in Beaumont as "Mr. and Mrs. Gladys City";

Dr. Joan Stiles, Professor of History, Lamar University, Beaumont, Texas;

Mrs. Nora Richardson, Sales Clerk, retired, Wolf-Wiles, Lexington, Kentucky;

PREFACE AND ACKNOWLEDGMENTS

Mr. Dan Corman, Corman's Decorating, Lexington, Kentucky;

Mr. Ryan Smith, Director of the Texas Energy Museum, Beaumont, Texas;

The talented Mr. Gary Christopher of the Christopher Photography Studios, Beaumont, Texas;

Charlotte Holliman, Special Documents, Lamar University, Beaumont, Texas;

Mrs. Jeanette Greer and Mrs. Helen Locke, Beaumont, Texas;

Mrs. Emily Jackson, a native Beaumonter, now of San Antonio;

Mr. John H. "Johnny" Walker, noted Beaumont historian and co-author of *Beaumont: A Pictorial History;*

A wonderful new friend, Mrs. Marie Broussard of Beaumont, Texas, who treated me, a stranger, so nicely when I visited Beaumont and who gave me such a royal tour of the city. What a generous ambassador of goodwill for Beaumont!

Barbara and James Broussard who invited me to my first "Spindletop Blowout

PASSIONS AND PREJUDICE

Celebration" in Beaumont, Texas, in 1996. That was grand!

Mr. Ralph Elders, Oral History Division, The Center for American History, The University of Texas at Austin, Austin, Texas;

Dr. John E. Gray, past President of Lamar University, and personal friend and business associate of Pansy Yount;

Mrs. Sally Blain, Secretary/Assistant to Miles Frank Yount and Pansy Yount, Beaumont, Texas;

Mr. Billy King, Supervisor of Magnolia Cemetery, Beaumont, Texas;

Dr. and Mrs. James Buie, Texas;

Mrs. Silas (Betty) Grant, Texas;

Mr. and Mrs. Jack Grant, brother of Cape Grant, Texas;

Mr. and Mrs. Fred Wachs, Jr., Lexington, Kentucky;

The P.R. experts at Joseph-Beth Booksellers in Lexington, Kentucky, especially Mrs. Peggy Collins and Suzanne; also my thanks go to

PREFACE AND ACKNOWLEDGMENTS

Mr. Wynn Morris and Mr. Hap Houlihan at Joseph-Beth in Lexington. What a great team of highly trained experts, and really nice people!

Photographer Cindy Nave;

The very skilled Mr. Danny Silvestri and the Impact Photography Studio staff, especially Dave, of Lexington, Kentucky;

Mr. Don Silvestri of Impact Photography;

Mrs. Francis Silvestri;

The very talented Mr. Sean Sears of Crystal Graphics in Lexington, Kentucky;

Mr. Jim Trammel of Crystal Graphics for his editing of the inside flaps of this book;

Mr. Carol Hale, artist and photographer;

Janie and Steve Taulbee of Taulbee Music for beautiful sound at my performances;

Mr. and Mrs. Lawrence Biggs and family;

Mr. Benny Biggs;

Rocella and Shannon Hurst;

PASSIONS AND PREJUDICE

My good friend, Rose Mansfield;

My precious sister, Patricia;

And the loving support of my dear mother, who always, faithfully without fail, asked, "How's the book coming, Linda?;" and who couldn't wait to see it in print.

I would also like to acknowledge prior fine articles and work done in this area that served as helpful background material written by Mr. Lynn Weatherman, editor, *The American Saddlebred Magazine*, Lexington, Kentucky; prior articles in *Saddle and Bridle Magazine* of St. Louis, Missouri; a master's thesis on "The Yount-Lee Oil Company" by Fred Barry McKinley, 1987, Lamar University, Beaumont, Texas; and the book *Spindletop*, an oil history, by James A. Clark and Michel T. Halbouty, 1952, 1980, and 1985, Gulf Publishing Company, Houston, Texas.

I sincerely thank *Saddle and Bridle Magazine*, and its fine editor Mr. Chris Thompson, for its outstanding article on this book written by a consummate writer, Joan Gilbert, in the September 1994 edition of the magazine; thank you, too, Joan, for not only your "Pulitzer"-caliber skill in writing, but also for your kindness and support in this project. My appreciation also goes

PREFACE AND ACKNOWLEDGMENTS

again to Chris Thompson for the *Saddle and Bridle* archive pictures used in this book.

I am indebted to Mr. Keith Bartz, Director of the American Saddle Horse Museum in Lexington, and also to their very capable Assistant Director, Ann Kraft, and archivist Anne Gabor at the American Saddle Horse Museum in Lexington, Kentucky, for their staunch support and for the pictures used in this book from the priceless collection of the American Saddle Horse Museum. Thank you all.

I am also beholden to Mr. Lynn Weatherman, editor of the *American Saddlebred Magazine* who was kind enough to offer additional pictures from his personal collection on Spindletop and its horses. Thank you, Lynn.

Other valuable source materials are acknowledged from the following newspapers:

The Lexington-Herald, Lexington, Kentucky;
The Beaumont Journal, Beaumont, Texas;
The Beaumont Enterprise, Beaumont, Texas;
The Louisville Courier-Journal, Louisville, Kentucky;

PASSIONS AND PREJUDICE

The San Antonio Times, San Antonio, Texas;

The Houston Post, Houston, Texas;
The Dallas Morning News, Dallas, Texas.

A special thank you must also go to the kind and patient people at the Jefferson County Court House in Beaumont, Texas, especially Joyce, Sharon Ned, and Sandy; the Tyrrell Public Library in Beaumont; the Orange County Court House in Orange, Texas; the Hill County Court House in Denton, Texas; the University of Texas at Austin's Center for American History, Oral History Division; and the American Saddle Horse Museum, Lexington, Kentucky.

I would also like to express my appreciation to all these fine folks who have supported my work and performances on Pansy Yount and Spindletop over the years:

Members of the University of Kentucky Woman's Club, especially Mrs. Muriel Varney and Mrs. Patricia Harris;

The Kentucky Humanities Council;

Members of the Lexington Lions Club, especially Bob and Ann Elsea;

PREFACE AND ACKNOWLEDGMENTS

Members of the Kentucky Bluegrass Area Extension Homemakers, especially Rita Hardman and Tina Peters;

Members of the Fayette County Homemakers Club;

Members of the Scott County Homemakers Club, especially Connie Minch and Jana Tylicki;

The Scott County Historical Society;

Mr. and Mrs. Reynolds Bell;

Residents of Richmond Place, Lexington, Kentucky, especially Mrs. Vi Smith who was the first person to exhibit faith in and actually support my movie project on this book. May God Bless Vi Smith!;

The Georgetown Rotary Club;

Mr. and Mrs. Grover "Deacon" Shropshire;

The Great Horseman, Mr. Frank Bradshaw;

Joan Gilbert, Writer/Author, Hallsville, Missouri;

PASSIONS AND PREJUDICE

The Woman's Club of Central Kentucky, especially Dorothy Manning;

The University of Kentucky Donovan Scholars Program, especially Roberta H. James and Mattie Umscheid;

JoAnn Smith and Betty Williams of the University Extension of the University of Kentucky;

Mrs. Kathleen McGraw;

The Burley Auction Warehouse Association, especially Denny Wilson and Diane Carlson;

The Campbell House Hotel of Lexington, especially Pat Holden;

The American Saddle Horse Museum's 1994, 1995, and 1996 "Saddlebreds in the Bluegrass" Tour Groups, the participants, and especially its talented Tour Director, Ann Kraft;

Mae Condon, *The American Saddlebred Magazine;*

Mr. and Mrs. Keith Bartz;

Mr. Lynn Weatherman;

PREFACE AND ACKNOWLEDGMENTS

All my Saddlebred friends, including Pam Hester of Magnolia, Texas--a real mover and shaker in the Saddlebred world; the Texas American Saddlebred Horse Association (TASHA) in Houston, Texas; Ada and Ed Perwien, Bluebonnet Farm, Bellville, Texas; Saddlebred trainer and horsewoman Sandy Currier; Mr. and Mrs. Malcolm Cole; my good and loyal friend Ann Kraft of Lexington, and on and on and on.

Mr. David Lubliner of the William Morris Agency in Los Angeles, who was working at the time for Richard Rosenberg, who first suggested to me over the phone that if I wanted to do a movie about this, I should really consider writing a book on it first. I remember how sincerely put off I was at the thought of sitting still for so long before a computer and putting this story in some form. It took me two very long years to realize Mr. Lubliner was right and to finally sit down and write this book. It was one of the best pieces of professional advice I've ever gotten.

To a crew member on the set of the movie *Sylvester*.

I would be remiss if I didn't thank, too, Mr. and Mrs. Martin Jurow for receiving me in their home in Dallas, Texas, and for their kind words. I carry a letter from Mr. Jurow around with me in my address book. Even today I look at it from time to

PASSIONS AND PREJUDICE

time because the encouragement of such a great and dedicated producer has meant the world to me--more than he will ever know. Thank you Mr. and Mrs. Jurow for your faith in this project.

I have been accused of trying to thank everyone but my kindergarten teacher. Let me just say, I would thank her, too, but I never went to kindergarten. The fact is, I am truly grateful to everyone who has helped me along the way!

P.S. Thank You most of all, God!

FOREWORD

The story of Spindletop and its lady is in many respects the *Gone With the Wind* of the American Saddlebred horse industry. It is the human story behind Spindletop and its lady--one of the greatest oil sagas the world has ever known. It is the personal, "inside story"--a story which has never been told before, presented from the inside out through the eyes of its lady, Pansy Yount, to vividly capture her struggles and the emotional storms that boil beneath the mysterious and haunting strata that is Spindletop.

It is a real-life, rags-to-riches story, a passionate love story, a story as big as its birthplace, Texas, and as colorful and glamorous as its adopted state, Kentucky--a story of Saddlebred horses and blue bloods, the clashing of iron wills, betrayal, and prejudice. It is the story of a strong woman--a woman who breaks out into her own rhythm late in life. It is in many respects our own story as struggling human beings--the good, the bad, and the ugly.

PASSIONS AND PREJUDICE

Spindletop's lady, Pansy Yount, was a business woman ahead of her times, a legend in the American Saddlebred industry, and founder of the first Saddlebred Horse Museum. She was a Catholic, benefactress to the University of Kentucky, the University of Texas, and Lamar University in Beaumont, Texas. She was a woman who helped charter a new course for world oil history, and who stood shoulder to shoulder with a man who became legend in the oil industry. Together, they put Texas on the map and made household words of the giants that grew from Spindletop oil, such as Gulf, Mobil, Texaco, and Sun Oil. A "Kentuckian" Kentucky will never forget, a Texan Texas can never forget, and an unconventional personality that rose above them all--she was Pansy Yount of Spindletop.

In real life, Texans know Spindletop as one of the greatest oil finds in history--one that propelled the world fully into the 20th century and created a new age for all mankind. In Beaumont, Texas, the memory of the first Spindletop oil gusher is honored for all to see by a monument-- the Lucas Monument--near the historical site of the oil gusher's first coming in, as well as celebrated by its own museum, the Spindletop/Gladys City Museum, near Lamar University. Too, there is a Spindletop Bash celebrating the anniversary of the first Spindletop oil gusher that takes place in January of selected

FOREWORD

years. The second Spindletop oil gusher, however, has to date never been celebrated, and the *personal* history behind it has never been explored. The first celebration in history of the second Spindletop oil gusher is planned for its anniversary date, November 13th, 1997, on its 72nd anniversary. At that time, Linda Light is scheduled to bring her one-woman show **Pansy Yount of Spindletop** to the Julie Rogers Theater in Beaumont, Texas.

According to noted Beaumont historian, Johnny Walker, "the second Spindletop oil gusher was perhaps economically greater in its impact than the first, since with the second, people knew what to do with so much oil, and not so much oil was lost when it came in." Mr. Walker speculates further that "if Frank Yount had lived and if Pansy Yount had stayed in Beaumont, there might have been a good chance that Beaumont would have been Houston instead of Houston."

The economic growth of Houston, Texas, was indeed profoundly impacted by Spindletop oil. This was due in part to the developing out of Houston ports to ship Spindletop oil out of Texas, to the locating of petroleum companies and headquarters in Houston, and to the proliferation of companies in Houston supplying equipment and materials to meet the ever increasing needs of the petroleum

PASSIONS AND PREJUDICE

industry. It is imaginable that such development could have happened to Beaumont ports and, thus, to Beaumont in time. Was such development just around the corner for Beaumont had Frank lived and Pansy stayed in Beaumont--if history had written itself just a little bit differently? It's an interesting question to which no one will ever know the answer. But Spindletop oil was, in many respects, the magic touch in the initial expansion and growth of greater Houston which gives Spindletop a type of special historical place of honor in both Houston and Texas history, as well as the history of Beaumont and the world.

Several universities have benefited from the generosity of Spindletop's lady, Pansy Yount: the University of Texas at Austin, for example, has been the recipient of a substantial amount of the Yount silver which can be seen there on display, as well as has received a bequest in her will and precious art works from the estate. Lamar University in Beaumont, Texas, has also been the recipient of some other Yount silver and memorabilia, thanks to Pansy Yount. Yet, Spindletop's lady is not to be found in books on Texas, even where Spindletop is featured. Her role in history has, up until this point, never been explored.

In Kentucky, Spindletop Hall on Spindletop Farms, just outside of Lexington, is one

FOREWORD

of the best-kept "*secrets*" in the Bluegrass; and the human story behind it is one of the most dramatically fascinating sagas in Kentucky history. Even today in Kentucky, the human story behind Spindletop Hall and Spindletop Farm is one of the most whispered about topics of curiosity and mystery behind closed doors and in social circles. Yet, Spindletop's counterparts in Kentucky, namely Spindletop Hall and Spindletop Farms, are to date totally without historical markers, and, to a great extent, rather closed off and "kept under wraps." They and their lady of mystery, with all they have meant to the state of Kentucky and to the world, are, for the most part, seemingly not remembered with public honors or publicly celebrated. This brings up questions for the curious.

People want to know the real story behind the rumors of Spindletop Hall and its lady. But, again, no one will find the name "Pansy Yount" in books on Kentucky, either. The glaring question, considering this woman's accomplishments and generosity to Kentucky, Lexington, and the world, is "*Why*"?

On the other hand, could it be that Pansy Yount has enjoyed a *secret fame* all these years that no one to date has put his finger on because it has been so masterfully disguised in a cloak of satire and sharp wit by one of history's most outrageous comic-strip writers? Could it be

XXVII

PASSIONS AND PREJUDICE

that Pansy Yount has had a faithful international audience of millions already moved "secretly" into place over a sixty-three year period through a celebrity role that neither she--nor anyone else, for that matter--would readily claim with pride as being based on themselves? Was a satirized *Pansy Yount* the "model-without-a-choice" for Al Capp's notorious character *Pansy Yokum* in *Li'l Abner*? A credible case can be built and argued that Pansy Yount may well have served as the inspiration, model, and cornerstone character behind Al Capp's biting satires against social hypocrisy in his broadly-drawn *Li'l Abner* between 1934 and 1976. It's an intriguing premise which also raises the questions of whether Capp's "Dogpatch" was a satirized Spindletop Hall and Spindletop Farms outside of Lexington, Kentucky, and whether Capp's "New York" was in reality a satirized Lexington "high-horse" society.

Spindletop Hall now serves as an elegant, private club for the faculty, staff, and alumni of the University of Kentucky, with leisure and sports activities, fine foods and drink, and posh dining. It is also a showplace and haven for couples dreaming of a special wedding or wedding reception in a prestigious location, rich in history, where one can literally step into the romantic beauty and mystery of times past.

FOREWORD

Today, Spindletop Hall and Spindletop Farms outside of Lexington are also prime location spots in Kentucky for shooting movies, both interior and exterior shots, with the colorful and very beautiful Kentucky Horse Park and American Saddle Horse Museum located on the same Iron Works Pike in Lexington. All three of these fabulous locations are within a stone's throw of one another. Portions of the classic movie *Black Beauty* were even shot at Spindletop in Lexington.

It was while working on one of the movies shot in part at Spindletop, *Sylvester,* a Columbia Pictures Production, with Martin Jurow of *Terms of Endearment* producing, that I first heard stories about Spindletop and its lady of mystery, Pansy Yount. As I listened that day with great fascination to the rumors, and gossip, and the stories of locals unfold--some true, most garnished with plenty of fiction--I could think of only one thing: "My God--they've got the wrong movie!"

The real movie (or television mini-series)--is within, in a powerful story that only life itself could write.

"She had a way of lavishing gifts upon people she hardly knew."
> Kathryn Manion Haider
> Granddaughter of Pansy Yount

"I recall how, on a shopping trip into town, Mrs. Yount stopped at a dime store for a few purchases and invited the entire sales staff out to her place for a Christmas party."
> J. Ed Parker
> Spindletop Farm Consultant

"Who does she think she is raising the Texas flag on Kentucky soil?"
> A Lexingtonian

"When I ride an American Saddlebred, I feel genuinely beautiful, inside and out."

Bonny M. Yount

PART I

Beginnings

BEGINNINGS

"That hill there has secrets. All kinds of secrets." Then the one-armed man would launch into one of his stories. With his most ghoulish voice, he'd tell about the ghost people said came down from the hill at night to Spindletop Springs by the river.

Our eyes would get as big as saucers as we huddled together under the blanket by the camp fire, like a covey of baby doves shivering with fear in the underbrush at the sound of an approaching hunter's footsteps, always keeping one eye on the one-armed man, and every once in a while wagering to cut our eyes on the sly toward that creepy hill, as if it were going to suddenly take legs and pounce on us and gobble us up mercilessly.

There were lots of superstitions surrounding the hill. Some said it was haunted. Some said spirits were guarding it. It stood ever-beckoning in the moonlight with its siren sights to the blind eyes and deaf ears that gawked at its mystery. There was a clump of closely standing oak trees on top of the hill. Heat rising from the sun-parched Texas earth around the cluster caused a constant wavy, shimmering motion, like a watery mirage that made the clump of oaks look like a giant "spindle top" that some mysterious force had

PASSIONS AND PREJUDICE

magically set in never-ending motion--ever spinning. Thus the name Spindletop Hill. And there were eerie lights that danced on the Hill at night. Patillo Higgins, the one-armed man, said there was a natural electrical field around the trees on that hill and that made the lights that danced like deadly teases of flirtatious lightning. But I was eight years old, and at eight I was sure they were spirits.

Patillo Higgins' one arm made him perfect for ghost stories. He was also a one-time bad boy who had seen the light and was now one of the Baptists' best Sunday school teachers. Some Sundays he would gather up poor kids from the neighborhood churches in Beaumont, and bring us on a picnic followed by an evening camp fire under the ever watchful eye of the Hill. I was Catholic, but that didn't matter. Baptists tried to save everyone, even Catholics, and the food was good.

If I got real close to the Hill, I had to hold my nose. It smelled like rotten eggs. But the sulfur and gases it exuded were perfume to Patillo's nose. He said these were earmarks, clues. "Oil," he said. "That hill is sayin' there is oil under it. Someday I'll be a millionaire and a good part of Beaumont with me."

The townspeople would laugh and poke fun at Patillo Higgins and his theories of oil on the Hill. He had no background in geology or oil. Geologists he'd approach called him "crazy." The

BEGINNINGS

people of Beaumont were only a little more polite. "The millionaire," they'd call him with a smirk. To his credit, his enthusiasm and belief did get three outfits to try to drill on the Hill, but they all three came up with dry holes. No one else would try for a very long time after that. But he maintained there was oil there and time would bear him out. And undauntedly he preached the gospel of the Hill according to Saint Patillo to all that would listen, for ten long years.

There my fascination began with Spindletop Hill--with ghost stories, the stench of rotten eggs, alluring dancing lights, and a one-armed prophet named Patillo Higgins. Though the years passed, I never forgot what I saw or what I heard that night: "That hill has secrets. All kinds of secrets." So began a saga that would change the face of the world forever.

PASSIONS AND PREJUDICE

I always thought if there was a hellhole, it had to be Orange, Texas. Not its people. They themselves are the jewels scattered now and again throughout its white-hot sands and steaming marsh lands, those watery mirages that fade in and fade out like wavy ghosts in the distance. But Orange itself has a type of desolate loneliness and hopelessness about it where everything seems worn down and out and old. Where loneliness is about the only thing you'll see comfortably sitting in the rockers on the deserted front porches that are too hot for even a lizard to appreciate, and where discouragement is like an ever-present, dull drone in the daily boredom the people call life. There is a pervasive emptiness that seems to drive people like herds of thirsty cattle the twenty-five miles to Beaumont toward civilization. Those that don't drive, run.

It was into this nothingness, with nothing, out of nothing that I was born, Pansy Burnadette Merritt, February 21, 1887. I was the youngest of nine children from a hardworking farm family, Hosea Holly Merritt and Sarah Francis Sherman Merritt. My mother died early on, and I was raised

BEGINNINGS

primarily by my oldest sister, Kathie Belle. My mother named me Pansy, Kathie Belle said, because at the end of the winter season every year, around mid-February, Mama would always try to set out colorful little pansies, but without much luck since Texas heat and dry soil aren't conducive to growing pansies. Pansies need dark, moist, rich soils and a comfortable, cool climate. So, Mama tried the real thing, and I was the little Pansy that came forth and into which she invested her hope, and would have invested her time if she had lived. Or that's the way I would like to believe it would have been.

We were poor, miserably poor. Seldom did I have a pair of shoes that fit, or a dress not too long for me or not made out of an old feed sack-- the kind with tiny faded flower patterns that we got down at the feed store when they discarded them. If we were lucky, we'd get more than one feed sack with the same pattern and not have to mix different patterns together when we sewed, just to have one whole piece of clothing. But I didn't know I was poor until I went to school.

All us kids would take a shortcut through Magnolia Cemetery in Beaumont on our way home after school. My mother was buried there; but there was no headstone. When the kids asked why Mama had no stone, I still remember that first searing, knife-like, almost guilt-laden pang of shame that shot through me, almost as if I had

PASSIONS AND PREJUDICE

done something wrong myself. And I'd lie, "We have one. But we ain't had time to put it up yet."

Then the kids would start laughing and taunting and chanting, "*Pansy, Pansy, here lies her mama's bones.... Pansy, Pansy, cain't ev'n afford an ol' gravestone. Pansy, Pansy, she's so dirt poor, look at those ol' feed sack drawers.*" Then they'd all laugh and run away. Many a day I spent sitting at my mother's unmarked grave alone and crying in the apron of my washed-out, coarse dress. And the giggling echoes of the children and the faces of their cruel delight never left me.

I was raised a strict Catholic. Every Sunday, Kathie Belle would make sure we were all cleaned up, and me and my brothers and sisters would all sit together in the front pew of St. Anthony's Catholic Church in Beaumont. Father would lead us in singing, and bless us, and for a while the pain inside of me would subside as I strained to imagine that everyone within the confines of those great doors practiced brotherly love; and, for a brief moment, I felt rich--at least in spirit.

I was never pretty in my life. Just average, if I worked at it.

BEGINNINGS

All of Beaumont was convinced Patillo Higgins' dogged persistence about oil being on Spindletop Hill meant he was losing his mind. His had been a voice in the wilderness now going on ten years. His theories were totally unsupported by any knowledgeable oil geologist and by any successful oil man anywhere. Over the years he had exhausted every possible oil outfit that could possibly consider drilling at Spindletop, except one, that of an old rugged seaman, Captain Anthony Lucas, who was game and now drilling at Patillo's behest on the Hill. Everyone stayed away from Patillo at all cost, because the minute he opened his mouth it was always about oil and Spindletop Hill. In short, the "millionaire" and anyone associated with him were to be pitied and avoided like lepers.

On the evening of January 9, 1901, the Hill seemed to give its most awesome light show ever, this time not only with dancing lights. For the first time ever, and the only time in its history, a great flaming ball of white light, which resembled huge balls of lighting that had suddenly joined hands, began to settle and explode over and huddle around the crown of Lucas' derrick atop the Hill. And a strange and great sound went up from the Hill like the hissing of what could have been the

PASSIONS AND PREJUDICE

earth itself's one and only great pressure valve. And the sound was heard for miles, like a trumpet heralding the end of one age and the beginning of another.

Then suddenly, on January 10, 1901, the laughing and pitying and leper-like avoidance of Patillo Higgins on the part of the townspeople stopped abruptly. The first Spindletop oil gusher was brought in by Captain Anthony Lucas; and Patillo Higgins, whose persistence matched the adversity he met, and who had held on to thirty-three-up-until-this-time-considered-worthless acres atop the Hill, became a real, live millionaire.

I was going on fourteen years old when the gusher came in. Wilder than the worst herd of mustangs stampeding, and louder than a freight train coming up out of the very core of the earth-- uncontrollable oil! It looked to my eyes like the oil was spouting over a foot wide and it was towering over the derrick, reaching to the blue of the sky itself. And at the top, if you followed it far enough, you could see the debris, and rock, and mud that had gotten in its way on its path out of captivity being crushed by the birthright of its power and rained back down on those around as if a fire-breathing dragon, once chained for centuries and cruelly treated, was now being unleashed and getting its first totally out of control taste of freedom and justice. There was the carnival jubilation of the townspeople who

BEGINNINGS

were gathering by the hundreds nearby for the sight. All these scenes were being indelibly printed with awe and reverence in my mind. It was like the whole world was dwarfed that day by Spindletop. Only Spindletop existed; and it seemed to enjoy its newfound royal reign over the past doubts of the people now gaping at it with open mouths.

It was a lesson not only in faith and persistence and the triumph of nature that I learned that day. I also observed the human lesson in how suddenly the town turned. Pitiful, one-armed Patillo was suddenly regarded with a type of quiet reverence, and the wisecrackers of just the day before were calling him "one of the wisest men on earth." He was at last a hallowed and respected native son in whom the whole town took pride; and, certainly, everyone claimed him as their best personal friend and, of course, they always knew he would succeed. Patillo was the last one in Beaumont to know of Spindletop's coming in, however, since, as fate would have it, this was the one day he was out of town. He had rode out early that morning to complete a deal in nearby Hardin County to help get him out of some serious debt.

When he came riding in, cold and exhausted that evening, he was hailed as a hometown hero, and a very rich one at that. From that day forth, debt was hardly a problem for humble, faithful Patillo.

PASSIONS AND PREJUDICE

The uncontrollable gushing of Spindletop lasted for nine days and kept the world on the edge of its seats. Spindletop could produce more oil in a day than all the other oil wells in the world put together! It was the beginning of a new age, the age of liquid fuel for the world. It caused people to sit up and take notice of Texas. Sleepy little Beaumont's face changed overnight from a model little town of cowpokes boasting 9,000 inhabitants to a Mecca for over 15,000 imported wheeler-dealers looking for their fortunes. Spindletop was producing over 100,000 barrels of oil a day!

There was another remarkable thing that happened on that historical day. My eyes fell on a young stranger in the crowd. He had blondish hair and piercing hazel eyes that were riveted on the majesty of the gusher. He was quiet and still and seemed to take focus over the hundreds of others that were gathering there in the crowd. He seemed more a part of the humility of the earth and the grandeur of the natural phenomenon being born out of Spindletop than a part of the crowd. He captured my attention just as the gusher had--by being its natural self. As we both

BEGINNINGS

stood there in the darkening shadow of Spindletop oil, there was an overriding sense of destiny and importance about both the young man and the Hill. And I wondered who he was.

PASSIONS AND PREJUDICE

I was eighteen years old now and went during this time mainly by my second name, Burnadette; I had met a man considerably older than I was, a successful oil drilling contractor, by the name of Charles Albert Daley in Saratoga, not far from Beaumont. He, with his experience, told me all the things a young, naive girl wanted to hear, and on Thursday, July 20, 1905, at 4:00 p.m., we were married in the Catholic church by Father.

Saratoga, Orange, Sour Lake, and Kountze were all tiny towns gathered like spokes of a wheel around Beaumont and all within a stone's throw of one another. We lived in two different places in Saratoga, and Albert had some money. He liked to smoke a big cigar and walk his bulldog around town in the evening with his diamond shirt studs gleaming. I had two bulldog pups of my own that I loved and kept in a fenced backyard. One evening when Albert and I were walking his dog together, we came back home late. I noticed my two dogs were lying sort of funny in the back yard and not getting up to greet me as usual. They always ran to greet me. Then I realized with horror the unthinkable--someone had shot the two pups through the head. I was heartbroken. I could not

BEGINNINGS

imagine someone being so cruel. I could not understand why anyone would be that mean to true-blue souls so pure, and innocent, and generous with their love. But someone was, and I cried myself to sleep over those pups. Maybe they were a symbol of Albert--or maybe of me--to some person out there. I never knew who killed them.

Albert and I were together less than ten years. It was an intolerably abusive situation for me. Albert had a way of making me feel subhuman, not even a member of the human race. He was callous and self-centered. I cleaned his house, made his food, and was there to be used whenever he felt amorous. He would come in drunk and sweaty and crawl all over me in his stupor; and then I would be alone for days, sometimes weeks, at a time as he got tired of the feelingless, cold sex he had with me and went out on me with someone more attractive, who for a little while satisfied his insatiable desires. When he finally came home, he'd graphically tell me about the sex he had with other women and laugh with that controlling, superior, what-can-you-do-about-it sneer. It was as if my very soul were being extinguished with his every word and every look. And I hated him for it. I hated him with a passion--a passion so deep that it was buried to me; I couldn't afford to feel the power of that hate. When I did, it scared me.

PASSIONS AND PREJUDICE

Sometimes Albert would start in on me when he was drunk for no reason at all. Drunks don't need reasons. He kept telling me how I was nothing, and he'd laugh and shake his head when I said I would leave if he hit me again. These were empty threats--I was too ashamed and beaten down to leave--that's what he knew all too well. People seemed blind to the actions of this fine, upstanding man in the community. He was a man without a blemish in some eyes, partly because of his small successes that brought him some money, partly because of the public face that he put on every time he left out the front door. Who would believe me? Many times I was so black and blue that I wouldn't even answer the door when someone came calling. Instead I'd play like no one was home and cower in a corner and pretend I was smaller than small and hope no one would barge in, in some friendly fit of thinking it was all right and that they knew me well enough to take that liberty. It got to where, if I did meet someone, my self-esteem was so low that I wouldn't look them in the eye for fear they could see what was really happening in my life--for fear that they would find out the secret of my painful existence behind what smile I could muster. And if perchance someone's eyes did meet mine, it would be almost startling and frightening to me, and I would quickly cast my eyes in a different direction, as if Albert's sins were my fault.

BEGINNINGS

Albert was so twisting my mind with his abuse that I was beginning to believe, in this god-awful situation, that something was wrong with me--that I was the problem. Albert, of course, had no problems. His pride assured him of that; and my nonexistent self-esteem was ever more shriveling up and concurring with that conclusion every time he struck me. I was lying to myself: Albert wouldn't be behaving like this if only I was a better wife. Being with Albert had made me sick inside somehow. Instead of running from him, the worse things got, the more I tried to please him--all so he wouldn't hurt me anymore, thinking that would make things all right. But that only made him more of a bully, as he tasted more control and the success of watching me flinch under his tyranny. I acted like an abused child, tiptoeing around, denying myself, begging for forgiveness when I had done nothing, desperately trying to salvage anything I could possibly twist around to resemble love. Even worse, I would deny what was going on, lying to myself that under all that coldness of his that was starving me to death and under all that cruelty was a shred of love worth saving. I became my own worst enemy, refusing to see things as they were, making excuses, and faking fleeting smiles just to survive. And I covered it all up in a neat package that I had been taught was religion, and made myself believe I was doing what God wanted me to do by "obeying" my husband, and sticking it out, and being faithful to the end.

PASSIONS AND PREJUDICE

I was petrified of Albert and his rages. And my hate grew as shame overwhelmed me. He'd always say, "You won't leave me. I have money and you have no place to go and no schooling. You don't want to cut your meal ticket off, now do ya, Pansy. What could you be but a prostitute, and then nobody would have you. Go ahead, I don't care about you, anyhow." He'd laugh that intolerable laugh again; and I'd cry out "Stop it! Stop it! Please stop it!" and break down as though shattered into a million pieces by a sledge hammer that just kept bearing down and down, harder and harder, breaking me into smaller and smaller pieces. He broke me until there was nothing left to break anymore. There was only a sliver of a thread left that held my soul--a thread that was fraying to a breaking point where no pride is left and the instinct to survive takes over.

One night when Albert was drunk, he hit me so hard that my face was bleeding. I was stunned like an animal that had just been hit by a club. He proceeded to have sex with me and then jerked me out of the bed when he was finished and dragged me to the door half naked--and with that ugly, horrifying sneer on his face, he threw me out into the yard like I was some worthless piece of dung. And he laughed and reeled as he closed the door. It was raining that night. I lay there in the mud, tears streaming down my face intermingling

BEGINNINGS

with the blood from the gashes Albert had caused with his fists, tasting my own blood as it spewed into my mouth from the cut above my lip left from Albert's ring imprint when he hit me, mud all over my bruised body. And Albert, in his stupor, opened the door again, with a big bucket of ice-cold water; and, as I tried to get up, he threw the bucket of cold water in my face, and I slipped in my attempt to stand. But he didn't stop there. He got another bucket, and another bucket, and another, mercilessly throwing it into my face until I was gasping for breath, almost like someone drowning in the force of a freezing river's current. And suddenly a rage that had been growing in me from almost ten years of injustice and mistreatment broke through my numbness and there was a moment of clarity--a moment of clarity arising out of all of the hatefulness, and disgust, and ugliness of what was happening that night. And I said and thought what I had never before dared to feel--"never again!!" The force with which I felt that rage frightened me. But I wasn't denying it anymore. It gave me the strength of fight that was life once more. And I began to crawl away in the mud, until I could get to a post and pull myself up. And then I stood, pulling myself together from the insides by that fragile sliver of a thread still abiding--still abiding and holding me in a singular piece, despite the pieces of me that Albert had broken over the years and strewn around him and trampled down. That frayed piece was suddenly, for the first time, gaining its own voice with ever

PASSIONS AND PREJUDICE

increasing fervor and determination. And, as if the recipient of an unlikely miracle, that piece of me began to retrace its steps. It began reweaving itself once again into a whole thread, then into a cord, and finally into a rope that I could hold onto. The sliver of myself that had reached its lowest point and was about to break was forging itself through the instinct of survival into a force to be reckoned with; and it was telling me that I was a human being--that I had rights, too--and that this was wrong!! It was enough!! Suddenly, that was all I could hear from deep inside of me--"ENOUGH!!"-- echoing like a welcome anesthetic to my pain. I couldn't think of anything else but that I didn't want this anymore. And I began ever so determinedly, still not sure on my feet, still somewhat reeling from the repulsiveness of this night, to walk away. And Albert, always with that nasty laugh, called with slurred words after me, "Where do you think you're going? Come back here! Nobody wants you. Nobody, hear me. You'll be back. And when you do, you're going to get it. You hear, you're going to get it!" And I kept walking away--away from Albert--in the mud and the rain into an unknown darkness that was more comforting--and I was not ashamed. For the first time in ten years, I was not ashamed. Not ashamed of being bloodied up, not ashamed of swollen eyes and bruises, not ashamed of clutching the few muddy scraps of cloth still left hanging from my mostly naked body. That night, I began the longest and most humbling walk of truth in my life-

BEGINNINGS

-the walk back to me--and my self-respect. Ashamed? No----no more.

I once heard someone say that sometimes we should be fucking the people we're fighting, and fighting the people we're fucking. It finally got through to me what they meant. Soaked, with blood streaming through the tears, groping in what was by now a cold, torrential downpour through the darkness, my fight came back.--And I began to find my way.

I walked all night long. I went through the woods to my sister Kathie Belle's in Beaumont. She was horrified when I appeared at the door. I don't think I even looked half human after all that had happened that night, and after all the strength that it had taken for me just to get to her steps. I had never shared any of my pain with her. Just imagine, she had actually thought we had a happy marriage, I had been so secretive. Somehow I hadn't known how to tell her. I was afraid and humiliated for so long. But she would have understood. And she would have believed me. She loved me.

The shame that I had carried was suffocating. It was literally killing me. That's what secrets can do. But that night, I gave the shame back to the man that had birthed it--back to Albert--it belonged to him. I wasn't going to carry it for him anymore.

PASSIONS AND PREJUDICE

As Kathie Belle wiped my face with a cold rag, I reached up to one of the gashes Albert had put near my temple. The blood I brought back before my swollen eyes looked dark, very dark-- more blue than red. As I flinched under the touch of the cold rag that Kathie Belle applied to my bruises and cuts, Kathie tried to distract me with soothing small talk. She noticed my squinting at the blood on my hand. She said the deeper the cut, the bluer the blood appeared at first sight, right before it started to flow. "*Blue blood* doesn't mean royalty and wealth like everyone thinks," Kathie said. "That's only its superficial meaning. If you're cut deep enough, your blood will look blue at first; and those deep cuts, if you can handle them right, can make you more noble than any coat of arms. How you handle your wounds-- that's what makes or breaks you. Determines your caliber of character." She dipped the rag in the basin next to the bed and folded it back gently, ever looking deeply into my eyes, into my soul. "Way down, everyone has a little blue blood to start with," she said. If you're lucky, and its refined by time, experience, and the school of hard knocks, that blood will become 'true-blue'-- worth more than any gold on earth." She kissed

BEGINNINGS

me and smiled faintly, feeling my pain as she touched my face. It was Kathie Belle's way of imparting a little wisdom. She was always more like the mother I had never known than my sister. I thought about Kathie's words--about what she said about "blue blood," and released myself, sinking into a deep, comforting sleep that carried me mercifully away from the night's agonies.

The bruises on my face and body began to heal, and the blood was no more. But the bruises and cuts on my soul would last longer. A shattered soul must be nurtured with the wisdom of an expert gardener who can give it care and light and help it heal--someone with the emotional and spiritual strength on reserve for two, who can carry you when you can't carry yourself. Kathie Belle was that gardener. There was a radiance and total unconditional and compassionate love about her that was willing to spend itself selflessly on all that she came into contact with. She was remarkable--my remarkable sister.

I gave myself time; it took time. I lived with Kathie Belle in Beaumont for a year and a half before I actually filed for a divorce on April 10, 1915. From the more detached distance that that year and a half gave me, I mused with Kathie Belle, after leaving Albert, that the "C" and "A" in Charles Albert's name hadn't stood for Charles and Albert at all, but for "cruel" and "adulterous." But I was religious, and that had held me hostage until I

PASSIONS AND PREJUDICE

suddenly realized that God was more than the rules of any man-made religion. I had a choice. And God was not the one condemning me for making that choice.

I filed:

Burnadette Daley	IN THE DISTRICT
vs.	COURT
C. Albert Daley	OF JEFFERSON
	COUNTY, TEXAS

To the Honorable Judge of said Court:

Burnadette Daley, plaintiff, complaining of C. Albert Daley, defendant, for cause of action, alleges:
1.
That plaintiff is an actual bona fide inhabitant of the state of Texas and has actually resided in Jefferson County twelve months before the filing of this suit, and that the defendant is a resident citizen of Jefferson County, Texas.

BEGINNINGS

2.
That a valid marriage subsists between plaintiff and defendant.
3.
That plaintiff and defendant no longer live together and their longer living together is wholly insupportable for the reason that the defendant committed various acts of adultery and cruelty.
4.
That there is no community property between the plaintiff and defendant and no children born of the marriage.
Premises considered, plaintiff prays that defendant be cited to appear and answer and upon hearing she have judgment dissolving the bonds of matrimony existing between them.

Attorneys for Plaintiff

On May 24th, 1915, I was freed from this insanity:

PASSIONS AND PREJUDICE

BURNADETTE DALEY	*IN THE COURT OF DOMESTIC RELATIONS*
vs.	
C. ALBERT DALEY	*FOR JEFFERSON COUNTY, TEXAS*

On this the <u>24th day</u> of <u>May 1915</u>, this cause coming on to be heard, the plaintiff appeared in person and by attorney, and the defendant...failed to appear and answer in this behalf, but wholly made default; whereupon <u>waived service</u> a jury being waived, plaintiff announced ready for trial, and the Court having heard the pleadings, evidence, and argument of counsel, is of the opinion that the material allegations in plaintiff's petition are true. It is therefore ordered adjudged and decreed by the Court that the bonds of matrimony heretofore existing between said plaintiff and defendant be and the same are hereby annulled and dissolved and that the said plaintiff be and <u>she</u> is hereby divorced from the said defendant.

It is further ordered by the Court that the said plaintiff <u>Burnadette Daley</u> do have and recover of the said defendant <u>C. Albert</u>

BEGINNINGS

<u>Daley</u> all costs in this behalf, for which <u>she</u> may have her execution.

<u>J.M. Conloy</u>,
Judge Presiding
60th District Court

As recorded in the divorce records, Albert didn't bother to show up for the court hearing and face me. I could now see him for what he was-- not a powerful man, but a coward in a bully's disguise.

I never mentioned Charles Albert Daley's name to anyone again. It was as though this part of my life had never existed for me or anyone else in my family. It was a chapter closed, and a secret sealed.

PART II

The 13th of Spindletop

THE 13TH OF SPINDLETOP

Things weren't easy. I took a job at Belle Knapps' Boarding House in Sour Lake to make ends meet. It was there that my eyes once again fell upon the stranger that I had seen that day so long ago, when we both stood in awe in the secret, dark mystery of Spindletop oil. The same blond hair, the same piercing hazel eyes--time seemed to have stood still for him. He came in every morning to Belle Knapps' and met with two other men at a corner table. He was quiet with a relaxed manner, yet had an intensity in his eyes-- strikingly handsome. I could sense a certain persistent determination behind his peaceful affability, much like the assurance ol' Patillo had years ago with Spindletop Hill. And he took focus in my eyes, just as he did that day when Spindletop first came in. We talked.

From that first moment when our eyes met, it was as though our souls were being joined in an ironclad union, a union of which neither one of us was willing to let go. We were soul mates, if ever there was such a thing, and seemed to understand one another with a warmth of communication that came from another more spiritual and eternal realm. And it began simply, quietly, effortlessly between us, as though the time had been appointed and had at last come. We were equals and kindred souls that had been, without our knowledge, somehow joined together in the depths of our being by the mystery and shadow of Spindletop oil, joined before a congregation that was all of Beaumont, joined ten years ago as we stood on the Hill in awe and silence and watched history being made. At this moment, however, some higher power, whose plan we would not have understood earlier and could not imagine even now, was choosing to let us in on and unravel the yet unrevealed secret of that hidden union. Suddenly, we found ourselves with willing abandon enveloped in an engulfing passion and ecstasy much like that which Spindletop itself must have felt before the whole world with its unleashed outpouring that day long ago, yet so present, showing each other our love with every touch of our bodies, as naturally as breathing in and breathing out, as if it had been written from the beginning of time.

No one could ever love me physically or emotionally as this man could. He was gentle, but forcefully masculine. And his hands--they were as wonderfully sensitive as he was. He always knew where and how to put his hands over my body in a way that made my whole body writhe in excitement. And I answered in kind. There was an electricity between us when we entwined our bodies and his eyes looked deeply into mine. He was like no other man in my life.

THE 13TH OF SPINDLETOP

This striking gift of God to me was Miles Frank Yount. He was thirty-five, and I was twenty-eight. Frank never called me Burnadette, but Pansy. My life changed, as did my name. Three and one-half months after I formally divorced Charles Albert Daley, I became the person I was meant to become, I believe, from the beginning of time. I became--Pansy Yount.

The Beaumont Daily Journal
September 16, 1915

Yount-Daley Wedding

A wedding which will be of interest to a large circle of friends in Beaumont and Sour Lake was solemnized at the home of the bride's sister, Mrs. Hardy S. Blanchette, 1408 Bibb Avenue, Wednesday morning, September 15, at 10 o'clock when Miss Pansy Burnadette Daley became the wife of Mr. Miles Frank Yount of Sour Lake.

According to the wishes of both parties the ceremony was very simple and dignified, Pastor E.P. Kennedy, pastor of the Central Presbyterian church, reading the impressive words which united their destinies: **"And the two shall become one flesh. What God has brought together, let no man put asunder."**

PASSIONS AND PREJUDICE

The wedding was witnessed by a very few intimate friends and relatives. The bride wore her going-away gown, a smart model in navy blue, with a "Bluebird" hat and accessories in harmony. Mr. and Mrs. Yount left on the 6:00 p.m. Frisco for the West where they will visit points of interest in Colorado, Washington, Oregon, and California.

They will be away until Christmas, after which they will make their home with Mr. and Mrs. H. S. Blanchette, 1408 Bibb Avenue, while they plan together the handsome new residence which Mr. Yount will have built for his bride.

The bride has made her home with her sister for more than a year and has made many friends during her stay in Beaumont.

Mr. Yount is a well-known oil producer of Sour Lake and is prominent in financial circles in Sour Lake and Beaumont.

Like most newspaper articles written by a well-intended third party, our wedding announcement was part truth, part fiction. True, I had lived with my sister more than a year; false, I had made many friends. I never had more than I could count on one hand in my whole life. True, Frank had come to Belle Knapps' Boarding House to meet with two men hoping to form an oil partnership with meaning; false, that he was a well-

THE 13TH OF SPINDLETOP

known oil producer as implied by the article; instead, he was struggling for success after some moderate successes as a wildcatter; true, that he was prominent in financial circles in Sour Lake and Beaumont, but not in the way one was led to believe by the article: it was two steps forward and three back; he went broke in his determination to find oil, if you call that "prominent in financial circles." And true, Frank wanted to build us a house that in our fantasies would be "handsome;" but that was hardly on the horizon, and we were still living with my sister and her husband in Beaumont. Also true, we did visit the West on our honeymoon, but Frank and I certainly scraped the bottom of the barrel to take that trip. I suppose you could give some poetic license to the article's author, however, and say the wedding announcement was basically correct. At least it was all right with us and in our minds and through the eyes of love; and, if not, it was certainly a bold and prophetic foreshadowing of the things yet to come, the things unseen in the spiritual realm that were seemingly on hold from the beginning, and which could only be brought forth into the realm of sight by a bulldoggish faith. Perhaps the greatest truth of the whole article were the words uttered by Pastor Kennedy that "united our destinies": namely, that Frank and I became one flesh. I believe we were indeed wedded in our very souls.

PASSIONS AND PREJUDICE

Frank and I came from similar backgrounds, but from opposite ends of the family spectrum. While I was pure Texan and the youngest of nine children, Frank was the oldest of thirteen children, born also to a hardworking, simple farm family named Nathaniel and Hattie Minerva Yount in Monticello, Arkansas, on January 31, 1880. That made him seven years older than I was. He was a devoted Presbyterian; and I, a devout Catholic. He took a lot of teasing about being a real razorback from hillbilly country, as folks called Arkansans. Arkansas was a part of the cotton belt. From the beginning, he was a self-taught mechanical genius and tinkered as a child with everything mechanical he could get his hands on. When he was nine he had to quit grade school in Monticello because his father died, and it fell to Frank as the oldest to help his mother with the farm and put food on the table for all his twelve other brothers and sisters. They were dirt poor. He always held a lot for education, partly because his formal education only went to the third grade. Hard farm work ate up his youth; and there was hardly time for any book learning, which he prized in stolen, secret moments. His dream was to be able to read someday, without halting or stumbling.

Miles Frank Yount was a steamroller in trousers. He had a dead-level way of locking eyes with people as if more comfortable with communing with their soul than at arm's distance through words. He was quiet, modest, and

THE 13TH OF SPINDLETOP

unassuming, and he was a dresser. More than anything, he liked to work. Work was Frank's joy in living. And he was ambitious. He left the family farm when he was eighteen and came to Beaumont and worked in irrigation in the rice fields, where he heard the stories of Spindletop and first became interested in oil drilling as a type of free-lancer, or wildcatter. As a matter of fact, his first drilling practice came as a water-well contractor for rice farmers. That's not to say he was good at everything, but a willing jack-of-all-trades who at first tasted a little of everything on Beaumont's banquet table of jobs while trying to find his niche. He would be the first to say he was a lousy real estate salesman and a lousy car salesman. He literally could not sell. But a mechanical genius he was, and a determined, persistent hard worker who didn't mind seventeen or twenty-hour days--whatever it took. Frank fed off of work. But in starting up in Beaumont he mostly contracted to thrash rice fields down with a crew he would hire, and then deliver the rice in bags to the farmer for market.

Frank had already put an irrigation ditch for one farmer all the way across one part of Spindletop, so he knew the Hill intimately. And his brother and him put all the money they could save together to lease a piece of land for a chance to wildcat on a small section of Spindletop. But as soon as Frank got the small well to producing a little, the farmer sold it out from under Frank for a

PASSIONS AND PREJUDICE

nice profit. Frank and his brother didn't know anything about legalities--and what the farmer had done was illegal--so they left the Hill without argument, defeated and disappointed, but Frank always dreamed he would come back--and he was always careful in legal matters after that. When we were alone, we both spoke and dreamed of Spindletop. It was as though we were caught in the unexplainably mysterious, alluring, hypnotic pull of the Hill. The Hill was a part of us, and we were a part of the Hill.

Where Frank went, I went. We were always pretty much side by side. Maybe I owed that to my father. Somehow he forgot to tell me along the way that a woman's place was in the home and that a woman could not do anything a man could do. He often took me around with him when he worked. Frank didn't seem to be aware of this widely-held belief of the times, either. To me, I was just being me; and it felt good and right. So, if we farmed, WE farmed; if we wildcatted, WE wildcatted. I was a late bloomer; and I had bloomed right into a young woman with mind and ideas of her own. This disposition was unnerving to some of the traditionalists, who believed a woman's place was quietly tucked away in the kitchen. I never noticed any difference in me and other women, however; I thought I was normal, and my tendency toward headstrong adventuresomeness was the salt and pepper of Frank's and my life together.

THE 13TH OF SPINDLETOP

Frank was no better off with his "steamroller in trousers" personality. He was constantly frustrated in attempts to form oil partnerships by the inaction and lack of vision of potential partners. A partner might have money for the operation, but no fresh vision; or a partner might like to act important, but not take a risk. To work with Frank, talk was cheap and you had to get a move on. Frank's intuition in oil matters tended to be more accurate than any scientific measurements; and once he decided on digging a well, he was a painstaking planner and tireless worker--a little like a bulldog with a steak bone between his teeth. There was no luck involved, but plenty of blood, sweat, and calm-as-a-cucumber cool when it came to Frank's successes in the field.

The two men that Frank was meeting every morning at the corner table of Belle Knapps' Boarding House when he and I first met were stockholders in the first Yount Oil Company that Frank had chartered already on February 13, 1909, and he was trying to stir them to action. Frank had leased a single acre on Spindletop from a former black waiter who had got the acre as a tip from another wildcatter who had had no success with it and was fleeing the Hill. The first hole Frank dug produced for a little while. But the second hole was dry and ate up all Frank's money; and the first Yount Oil Company went under--completely bankrupt. It was as though Spindletop was trying

PASSIONS AND PREJUDICE

to test Frank's grit with a slap in the face, and thus test his worthiness.

Frank tried to get two other partners to buy a small track of land in Sour Lake. But when they learned it was surrounded by dry holes already drilled by the Texas Oil Company and that Frank was operating on a "hunch," they turned him down flat.

That's the point at which Frank became perturbed and tried to sell cars and real estate, but that didn't work. Frank's heart and talent was in oil.

Frank once again tried to buy a tract of land through partners, again in Sour Lake on another "hunch;" but no one could see what Frank saw with his sixth sense for oil, and the partners were again afraid to take the risk with no physical evidence before their blind eyes.

That's when he met a man of action named T. P. Lee, and at last a partnership was formed that would produce: the Yount-Lee Oil Company. It worked mainly because T. P. was smart enough to recognize a natural genius and smart enough to listen and not dismiss Frank's intuition and hunches, thus giving him free reign. Frank was the Yount-Lee Oil Company! And on April 13, 1922, the Yount-Lee Oil Company hit the first great flank oil well on the Gulf Coast.

THE 13TH OF SPINDLETOP

But no matter where Frank managed to bring in a producing well, his eyes were always on Spindletop. There was still that personal challenge and attraction between Frank and the Hill that had been informally but firmly extended when it sunk him once years before. It had become a matter of honor.

But life writes its own scenarios. Something happened. By 1925 the Hill suddenly became a forgotten shadow, a ghost of itself. The Hill that no one ever saw an end to, but only a perpetual beginning for, petered out. Too sad an end to the glory Spindletop had known! Gone were the "Diamond" Jim Bradys and the Andrew Mellons who made their fortunes from the black, oily blood they drained from Spindletop's bounty. Gone were the train loads of hopefuls who had poured into Beaumont looking for their fortunes. Gone were the rugged, loud oil riggers and the carnival-style wheeler-dealers who cut colorful deals at every corner. Gone were the first bloom and rosy blush from the cheeks of a young booming Beaumont. The Hill was seldom spoken of anymore. It had been ravaged, raped, and left for dead without so much as a marker to signify it had ever existed.

There were times when Frank and I would go out to the Hill and just stand and look at its lonely desolation, with its sandy, dry, hot grains of dirt being whirled up by some equally hot and equally

PASSIONS AND PREJUDICE

lonely small wind who was just looking for a friend to pass the time with. Sometimes it seemed that the Hill was trying to tell us something with soulful groanings as excruciating and frustrating as an animal that had had its tongue cut out with senseless, savage disregard. As we looked at Spindletop, there was a hurting deep down inside of us, and a knotting in our throats. The Hill was barren.

The Hill's condition seemed to bring up a sadness and void inside of us that Frank and I had shared in painful secrecy and seldom talked about since we married. We could not have children. That was something we wanted more than anything, but all the money and success, and hard work, and oil wells in the world could not buy.

There was a big epidemic that struck the Sour Lake oil fields in the 1920's, and my best friend called me one day and said, "Pansy, come quick." Her husband had just died in the epidemic; and she herself was stricken and in so bad a shape that they didn't know if she would make it. Hundreds had died in the camps of the oil fields. Frank and I went immediately. But it was not quick enough. When we got to the Sour Lake fields that early Texas morning, my friend was dead; and beside her was their newborn baby, a little girl, crying at the top of her lungs, alone, wrapped in her little blanket beside her dead mother's body. Frank and I looked down at the child who

THE 13TH OF SPINDLETOP

extended her tiny hands and arms toward us; and we picked her up in our arms and held her tightly. As she looked into our eyes, and felt the warmth and security of our hands around her tiny body, her little face suddenly unwadded from its contortions. There was only that little helpless sniffle of a whimper left in her, as the tears still ran down her red cheeks, and she gasped to catch her breath after all of the frightening flailing and crying at feeling the unexplainable desperation of abandonment in death. Then Frank's eyes met mine. We adopted that little baby girl, and took her home to raise as our own. From that day forth she could not have been more our own than if she had been our flesh and blood. We loved that child with a passion. She was the apple of Frank's eye. That was May 2, 1920, and we named her Mildred Frank Yount after her father, Miles Frank Yount.

Despite what appeared to be the outward condition of Spindletop by 1925, Frank had never lost faith in the Hill. He saw something that other eyes were blind to. And he did not want to lose the Hill's challenge to him years ago when the Hill bankrupted him; and he certainly did not want to win by default. That would not have been honorable in Frank's eyes. He felt he was worthy of the Hill's challenge, and the Hill was worthy of another try. Mildred was now five years old.

PASSIONS AND PREJUDICE

It was then that I heard of a man named Marrs McLean who still had some oil leases on Spindletop Hill and wanted someone to take them over and work them. He, too, thought that the Hill was worth another try. He had gone around the world looking for someone to take over his oil leases and had now returned to Beaumont and his office in the San Jacinto Building, which just happened to be next door to Frank's Yount-Lee Company office. (Frank had moved the oil company headquarters from Sour Lake to Beaumont and the San Jacinto Building in 1923.) So there sat Frank with his passion for Spindletop on one side of the wall that separated his office from that of Marrs McLean, and Marrs sat on the other side of the wall in his office with the oil leases for Spindletop still in his lap, wondering if he'd ever find someone willing to give Spindletop another try.

I told Frank what I had heard about Marrs, and the very next day Frank and I walked into Marrs' office. Oddly enough we had known Marrs before all this; but, for whatever reason, Frank never thought of Marrs, and Marrs, right beside him, never thought of Frank. It had simply never occurred to either man to contact the other next door. When Frank asked him why he had not offered the Spindletop leases to him, Marrs said that he really didn't know, but he guessed he had thought it would take a major company to handle the deal. No one really knew the name Miles Frank

THE 13TH OF SPINDLETOP

Yount that well in Beaumont. Marrs asked him then and there if he wanted the Spindletop leases. Frank did not hesitate to accept.

Both were ecstatic at their good fortune of finding each other; and I was equally happy at the match. Frank took over Marrs' oil leases on Spindletop with only one condition being placed on the contract from Marrs: that Frank first drill where Marrs thought there might still be oil. It was Marrs' pet idea. If there was no oil at that spot, Frank could then have free reign over the leases and drill wherever he wanted, as long as he wanted, and in any way he wanted. It was May 11, 1925, and Frank was relieved to have at long last signed the leases to Spindletop Hill.

Frank was true to his word. He drilled exactly according to Marrs' first-choice location on Spindletop Hill, but came up with a dry whole.

It was a little like rolling dice. Now it was Frank's turn. He knew exactly and precisely to the inch where he wanted to drill: a quarter of a mile from the first Spindletop Gusher, and only twenty feet from the first place Frank ever drilled on Spindletop Hill--the one the old farmer had sold out from under Frank and his brother illegally years ago. He had had to wait for almost fifteen years since that incident. He had studied the Hill and its moodiness for years. The drilling was now a well-calculated, well-planned, intuitive hunch guided by

PASSIONS AND PREJUDICE

the solid, sure-footed "insides" of an experienced oil driller whose track record of successful hunches had long bypassed the best scientific calculations geologists had to offer. The hole now was being drilled by a Miles Frank Yount who had learned through the school of hard knocks. Frank had earned through the exhausting, hard work of the years the right to be called a self-made oil man. The little-educated, small Arkansas farm boy, who had tinkered with all that was mechanical, was now the Texas oil man drilling with more than book knowledge and with full steam ahead.

Hours were never counted by Frank or his oil crew. As with all worthwhile endeavors that a person is meant to do, there was no passage of time.

I worked alongside Frank many a night, giving light to him and his crew with a second-hand lantern I held. That was all we could afford at the beginning. Other times, I filled hard-working, empty stomachs with food, serving as the rig's on-site cook. I had always liked to cook.

In the evenings, Frank and I would nestle in each other's arms before the fireplace in our modest cabin and share our innermost dreams and thoughts with one another. We'd stare into the fire as it crackled and danced with curious images on the cabin walls; and we'd kiss one another and

THE 13TH OF SPINDLETOP

make love in the profound security that we felt in each other's arms. All the day's work and all the cares of the world would fade into the soothing murmurs and touches that were the love and passion that Frank and I shared together. We had a once-in-a-lifetime love, the strength of which seemed to override whatever obstacles might appear before us.

At other times, Frank and I talked a lot about drilling. He had been the first person to ever attempt using a rotary drill in Texas, a drill that would drill deeper into the earth's hard rock than other drills. Frank had the idea that he would drill not from the top of the Hill like others had, but on the south flank; and he would use a rotary drill and drill deeper than others had ever been able to or were willing to wager. He reasoned that the Hill was like a plate of glass. If you took a hammer and hit the glass plate, all types of breaks would extend from the place you hit. In Frank's mind, Spindletop Hill was like that plate of glass that had been hit: it must have underground veins extending in all types of directions. And there was oil in everyone of them, including those running down the flanks, he thought.

It was November 13th, 1925, the night of the Southeast Texas Fair. The fair was huge, and all of Beaumont seemed to be there that night. They had a real show-off, "cutter" of a fiddler playing with a square dance band. The caller was leading square dances, and Beaumont seemed to be enjoying itself whooping and hollering under the spectacular fireworks display they were having, and dancing that night under that big Texas moon. Somehow I always thought nothing could ever be bigger and prettier than a full moon in Texas. The sky always seemed bluer, too; and the stars more plentiful. But I was partial; my roots were deep, and I loved Texas. It was an inseparable part of me, and I had never known anything else.

Frank and I had stopped by the fair that night along with everybody else. When we heard the band strike up *San Antonio Rose*, I dragged a reluctant Frank onto the dance floor, and together we made the world go away in each other's arms. It was like it was only the two of us on that crowded dance floor that night. The melody of *San Antonio Rose* when carried by an expert Texas fiddler has a hauntingness about it--a feeling of strength and determination in its simplicity, yet a lingering, stubborn loneliness that grabs at you from within its strains. And when Frank and I held one another within each other's eyes and moved together that night, it was like that song and the fiddler's carrying of its haunting melody framed the way Frank and I felt about each other under that big Texas moon. Afterward, as we walked through the endless fair booths with cotton candy in hand and with the barkers clamoring at us, Frank stopped long enough to win me a small stuffed horse by throwing balls at weighted wooded milk

THE 13TH OF SPINDLETOP

bottles and knocking them down. I prized my little make-believe horse, kissed Frank, and then caught him off guard momentarily, long enough to playfully try to smear his nose and mouth with a finger full of cotton candy. As he tried to wipe it off his face with his handkerchief, he then tried to steal some of my candy and get me; but I was fleet of foot to dodge when I saw him coming. We were happy together that night.

Frank spotted a young fair artist close by, and he asked him to do some caricatures of us both. Frank thought that would be fun. It only took a few minutes for the young man to do some pen and ink sketches. He did mine first; I never liked any picture of me all that much. You can't turn a sow's ear into a silk purse, as the saying goes. But it was a good try by the young man. Then Frank posed for his. I was standing a little over to the side watching when a man came up from behind me and grabbed me by the arm so tightly that it hurt. Then the voice came deep and throaty and almost as a husky whisper into my ear, "Well, what do you know. Pansy. Small world, ain't it?"

As I turned, startled, protesting loudly with "let go of me," I saw his eyes and the tobacco he was chewing oozing from between his teeth, and his dirty, unkempt look. Frank glanced up, bolted, swung the man around, and laid a good-sized,

PASSIONS AND PREJUDICE

power-filled fist up against the man's head, sending him stumbling backward.

"Don't ever touch my wife." Frank's voice had a calm, low deadliness about it that no one in his right mind would have questioned. But the man was feeling no pain, except where Frank had launched into him and blood was starting to come. He stumbled back up.

"Wife?" Then he laughed, "Pansy?" Then to his buddy, "Why, she and I go way back;" then to me, "Don't we, Pansy?" Then he got louder. "She goes way back with a lot of men up around Austin way, don't you, Pansy? Matter of fact," the man kept on railing, "she ain't nothing but --"

He was clipped short by Frank grabbing him by his collar and shoving his gun under his throat so tightly that the man was choked for air.

Frank was right in the man's beer-reeking face, gritting his teeth, and in his controlled way told the man to "shut up!" Then he shoved him hard into his buddy, who was quick to try to smooth things over.

"I'll take him, Mister. He don't mean no harm. He's just had a little too much, that's all." Then he pushed his buddy, who was still somewhat disoriented, wiping blood from the side of his head, into the crowd, and the two disappeared

THE 13TH OF SPINDLETOP

into the nowhere from which they came. A crowd had gathered and were watching the commotion with curiosity. Ladies were whispering; and some of the men had a knowing grin turned up in the corners of their mouths, which they quickly dropped when Frank's eyes met theirs. Frank was normally a very calm, peaceful man--but not where I was concerned. He loved me and protected me at every corner.

I looked at Frank, and he at me. Then he took my arm gently but firmly and said under his breath with strength as he eyed those watching, "Come on." He bothered to quickly but quietly pay the young artist and take our sketches with him; and we briskly left toward another part of the fair. We walked side by side. Frank's eyes looked straight ahead. We didn't say a word for a long while, but just walked. There was nothing to say. What that man had said did not come as news to Frank. There had been rumors before. Maybe they had come from Albert. I'd never know...but that was all a long time ago.

Frank slowed up after a while, and put his arm around me. He turned to me under that big Texas moon that night, and looked into my eyes with his intense hazel eyes, then kissed me, and held me with a strength of love and acceptance that I had never known before from any other human being in my life. Frank loved me with a passion--there was never any question in my mind

PASSIONS AND PREJUDICE

about that. We were more than husband and wife--we were friends, companions--and we trusted one another as equals. He managed a smile, then paused, and said, "I want to show you something. Over here."

They were having the usual bronc and bull-riding rodeo over to one side, and a special horse show over to the other side, featuring something they were calling "American Saddlebred" Show Horses. I had always loved horses, but had no idea what Saddlebreds were, certainly something not all that common in Texas, where rodeos and bronc riding occupied center stage. It piqued my curiosity. Frank knew about them though, and took me over to watch them do their stuff. And what stuff it was!

I had never witnessed anything so beautiful before. These were not ordinary horses, they were artists. Every muscle and vein of their massive, alert bodies and every step and turn they took was sheer poetry in motion in the most elegant display of gracious power, skill, and beauty I had ever seen. Totally captivating, breathtaking; I could not take my eyes off their every move. I felt like my heart skipped a beat as I watched them, and I fell in love that very moment. I turned to Frank and told him that if ever we had enough money I wanted to go into show horses, Saddlebreds. They sure beat the hell out of rough-and-scrubby bucking horses in my eyes.

THE 13TH OF SPINDLETOP

I remember the way he looked at me when I said that under that big Texas moon that night. He smiled. He took the little stuffed horse he had won for me earlier, and, looking at it and me, said, "Maybe this will grow into a real live Saddlebred someday, who knows?"

"You think it's some kind of sign?" I replied, smiling a little at his tease.

"You never know." And then as we walked further, arm in arm, "But I promise you one thing: if ever we have the money, that'll be one of the first things on our list." And he kissed me. "We'll do it," he said. "We'll sure do it." And he squeezed me tight as he kissed me again. We walked along in silence for awhile. Then Frank said if I ever really wanted to go into the American Saddlebred business though, there was only one place to be-- in the Bluegrass of Kentucky.

I thought I had heard the name Kentucky somewhere before; it sounded familiar, but I had no idea where it was. It just as well have been a foreign country; I had never been over a hundred miles from Beaumont in my entire life. I asked Frank if the grass there was really blue. He said only in the early morning dew, and smiled again. The beauty and majesty that night of those most perfect animals, the Saddlebreds, performing before the Texas crowd, never left me; and I mused to myself about the strange "blue" grass

PASSIONS AND PREJUDICE

that grew somewhere in a place called Kentucky. I thought this "bluegrass" must be some peculiar food that only Kentucky grew that must make Saddlebred show horses the most beautiful in the world.

From that day forward, my hidden, innermost desire and fantasy was to go into Saddlebred Show Horses. I even dreamed of it at night.

This November 13th was not only the night of the Southeast Texas Fair with its colorful excitement and unexpected encounters. Frank was superstitious and partial to the 13th. He had first organized his Yount Oil Company on February 13, 1909, and he brought in his first real producing well on the Gulf Coast on April 13, 1922. And here it was November 13th again. Frank had one of his strongest intuitive urgings and "hunches" on this day. He and his oil field manager had taken a sample of the oil sand and a bottle of ether earlier in the day and gone to his office to test it. They both agreed they "might have a wild Indian to fight" soon. In a quiet and confident manner he told a few people he might have something special for them to see if they wanted to meet him over on the Hill later that evening. But no one listened to Frank and his idea that there was still a second wind in Spindletop anymore than they had listened to Patillo Higgins or Captain Anthony Lucas who had brought in the first gusher. Men in oil circles didn't really know Frank Yount and could have

THE 13TH OF SPINDLETOP

cared less. Frank had now been drilling steadily and persistently, as was his nature, in the same place for six months using a rotary drill; and he was at the unheard-of depth of over 9,000 feet.

After stopping by the fair that night, Frank and I went over to the Hill. It was only a hop and a skip away, so the music from the square dancing and the caller was coming in loud and clear to all the crew still drilling. I remember Frank taking off his jacket and rolling up his sleeves under that flashlight-clear, full moon, and joining his grease-laden men at the derrick. I gave them additional light with the second-hand lantern I always brought with us. It was all we had.

People say that if a person has a dog long enough, he starts to resemble the dog, or vice versa. Maybe if a man drills an oil well long enough, the well takes on the man's nature. That night the nature of the Hill and the nature of its would-be master merged.

Suddenly Frank's derrick man who rode point up on the derrick leaped off just as the whole derrick splintered, and shouted that cherished phrase, "Coming out of the hole." Everyone scattered. We were all stunned and frozen for a split second, confused both by the unrealness of what we were hearing and by the abruptness. Could it be?! Yes. Spindletop was coming back!! And it came back in with a roar from the core of

PASSIONS AND PREJUDICE

the earth that shook the fairgrounds like an earthquake and turned all eyes in its direction. But unlike the wild, uncontrollable, first gusher where hundreds of thousands of barrels of oil were lost during its first uncontrollable nine days, this time Frank had taken the precaution of installing a control pressure valve, just in case. So Spindletop came in under perfect control; not a drop of crude was lost at the experienced hands of Miles Frank Yount. Whereas Spindletop's first arrival had been noisily trumpeted from the depths of the earth, this arrival came with the quiet confidence of one who had already been there and was familiar with the way things were and was equally hard to impress, much like the quiet, self-assured man whose faith and grit the Hill had tested by fire through the years and finally found worthy of its secret. We were covered in oil. At last, Miles Frank Yount was master of the Hill.

THE HOUSTON POST

AN ELECTRIFYING ANNOUNCEMENT AT THE STATE FAIR: COMEBACK

About 8 o'clock the evening of November 13, thousands of people at Beaumont's South Texas State Fair received an electrifying announcement over Magnolia Petroleum Company's new experimental radio station.

THE 13TH OF SPINDLETOP

"Spindletop has come back," John W. Newton, the announcer, said excitedly. "The Yount-Lee Oil Company's Number 2 McFaddin on the south flank of Spindletop dome came in not more than an hour ago flowing an estimated 5,000 barrels of oil a day through a small choke."

The fairgrounds emptied almost immediately and repopulated the fields surrounding Spindletop with company officials, friends, office workers, hundreds of curiosity seekers, oil scouts, newspaper reporters, and thousands of other people who had thought that the Southeast Texas Fair was going to be the only attraction of the evening. Frank remarked to one person there with his usual optimistic, quiet confidence, "These flanks will produce another 60 million barrels of oil." The field under that clear Texas moon that night was a sty of mud, oil, and jubilant people. And the Beaumont newspapers read:

But there was no one in the crowd as jubilant as Pansy Yount.

There was never a truer statement written.

PASSIONS AND PREJUDICE

I didn't know it at the time, but it was the beginning of a new age for the world. The first gusher had changed the face of Beaumont and given rise to an expanded Port of Houston, making people sit up and take notice of Texas as it shifted its focus more from cattle and agriculture to oil. But the second Spindletop solidified Beaumont's future and gave Houston an economic shot in the arm that made it a hard act to follow from that day forth in international business and economy, as Houston ports became larger and larger, Houston petroleum-based businesses and suppliers flourished, and Houston's development became more and more sprawling. Thus, the second Spindletop oil gusher put Texas *solidly* on the economic map. This Spindletop brought in twice as much oil as the first and ensured a new course for world oil history. And whereas the first gusher began to change the character of civilization, the second continued and stamped this change permanently onto world history, propelling human progress at least a quarter of a century into the future. Spindletop changed the lives of common people everywhere. As one article put it:

> **The wheels were speeded up and the internal combustion engine developed. Men like Henry Ford and others took the individual off their feet and put them on wheels and wings to travel the world. The most valuable product of crude oil until then was**

Another Premium Product

At No Extra Cost!

Sun Oil Company

San Jacinto Bldg. Beaumont

SUNOCO
Heavier Bodied Oil

PASSIONS AND PREJUDICE

> *kerosene. Now hundreds of items valuable to human lives were going to be made of petroleum.*

Fortunes on the magnitude of those of Howard Hughes and J.D. Rockefeller would be brought forth out of Spindletop Oil before all was said and done. Oil giants such as Texaco, Gulf, Crescent, Magnolia, and Sun Oil would rise out of this abundant source. Even Rockefeller's Standard Oil would receive a giant shot in the arm from Spindletop. It was one of the greatest oil wells ever discovered in the world, second only to one found in Russia some years before, which was now dry.

And it was a new age for Frank and me. I was thirty-eight and Frank was forty-five when Spindletop came back in. Our whole lives changed on a dime that November 13th. It was as though our wedding announcement of ten years ago had foreshadowed this historical rebirth. Our modest frame house suddenly turned into that "handsome house" that Frank Yount planned to build for his bride--an opulent mansion that we bought at 1376 Calder Avenue in Beaumont. We called it "El Ocaso," Spanish for "The Sunset;" and we built a picture-perfect summer house just outside Colorado Springs, Colorado, in Manitou, which we called "Rockledge" that was straight out of a fairy-

THE 13TH OF SPINDLETOP

tale, complete with amusement park. Our wedding announcement had called Frank a "successful oil producer in Sour Lake" who was, tongue in cheek, "well known in financial circles." At the time, considering the Hill's once bankrupting Frank, that was a kind overstatement. Now Frank was heralded by the *Beaumont Enterprise* newspaper and newspapers nationwide on a totally different scale:

M. F. YOUNT OF BEAUMONT IS COUNTRY'S MOST SUCCESSFUL PRODUCER OF PETROLEUM

We were suddenly well known in financial circles for more than Spindletop having bankrupted Frank years ago. Although Frank and I never revealed our worth during the Spindletop years, we were more than millionaires--in today's comparable dollars we would have been billionaires. And people who, up until Spindletop came back, would not even give Frank and I the time of day suddenly became our new friends, hanging off of trees with politeness and hellos. I would say in reality we had only as many real friends as we had ever had--less than the fingers on one hand.

PASSIONS AND PREJUDICE

The voices of my schoolmates and ghosts of my past were still noisy, vivid, and ever-present in my head, like guilt-ridden children hiding behind a curtain and taking some unknown pleasure in compulsively writing on their mother's clean walls. Do you know what I did with that first money from Spindletop? I bought my mama a proper gravestone for where she lay in Magnolia Cemetery. Yes, I made sure, at last, that the laughter stopped and my family had proper gravestones. They deserved at least that.

Frank and I took the whole commotion that now surrounded us with a certain healthy sense of humor, mainly because we had been without for so long; it seemed unreal. We were still us. Spindletop changed more the things and people around us than it changed us. We still had the same insides as when we were both growing up dirt poor. Frank only wanted to work; and my life beside Frank was everything I wanted it to be, plus an added measure of adventure that appealed to me.

Frank didn't drop a stitch. He immediately drilled a second hole on Spindletop Hill, and that came in strong, too. That made it an even more unlikely "two in a row." The date, again the magical 13th--January 13, 1926. From there he went on to build the biggest oil-tank farm in the world south of Spindletop across a canal he once worked on as a ditch digger. The project was started on none

THE 13TH OF SPINDLETOP

other than August 13, 1932. So played the mysterious 13th in Frank's life.

Frank was full of surprises.

He said it was time to start building a barn. Frank had never forgotten that night at the Southeast Texas Fair and how awe struck I had been at those Saddlebred show horses and the name Kentucky. He remembered my words to him that night under that big Texas moon: "If ever we have money, I want to go into Saddlebred show horses. They sure beat the hell out of scrubby rodeo broncs." He had just gone up by Dallas to the Mesquite Rodeo and announced that he had run into and hired on the spot a special Saddlebred horse trainer--the best in the business--and given him a blank check with orders to go out and find the best and most beautiful American Saddlebred show horse prospects money could buy and bring them back to Beaumont. We were going to start a Saddlebred stables, and the barn would be ready by the time the trainer arrived in Beaumont with the horses. I was ecstatic and deeply touched by his thoughtfulness and memory for what was important to me. This was just one of the reasons I loved Miles Frank Yount, because it was obvious in everything he did that he loved me and

PASSIONS AND PREJUDICE

put my welfare first. And he never broke a promise. Construction began, and thus was the beginning of our eighty-acre Spindletop Stables in Beaumont, just to the side of our home on Calder. The trainer's name: William Capers Grant, or "Cape" Grant as he was known in horse circles.

Meanwhile, Frank hired himself a tutor and got himself a set of McGuffey readers and set about to learn to read. He was determined to take up where he had left off in schooling and become not just a reader, but a good reader. And Frank did just that with determined hour after hour work, sometimes lasting late into the night. Sometimes I would go past his study in our house on Calder, and he'd be sitting there, reading aloud, struggling with a new reading assignment, sometimes slow, sometimes halting, but ever steadily forward and smoother. He worked, and he worked, and he worked until he conquered the old ghost and secret pain that had been haunting his life up until then--plain ol' reading. Then he took a room in our house on Calder and turned it into a beautiful, warm, wood library and populated it with books on every conceivable subject, all of which he read.

Mildred was indeed the apple of Frank's eye. He laid a lot of worth on her education. Frank said he wanted to give Mildred Stradivari violin lessons, and to that end got every book he could lay his hands on on Stradivari violins and read them so he

THE 13TH OF SPINDLETOP

could be knowledgeable about the violins and talk intelligently about them. And he got Mildred the best Stradivari violin lessons money could buy and even took lessons along with her to encourage her. An encouragement he was; God knows he could barely squeak along. But Mildred played like an angel. He got her a Stradivari with her name on it, and one for him with his name on it. They were two inseparable peas in a pod, Mildred and Frank. And he loved to hear her play in the evenings.

It didn't take much to make headlines. It was as though we were bugs under a glass being studied by the media: every step we took seemed to show up in the papers. This never sat well with Frank. He was by nature a shy, generous, modest man who loved his family and loved working, but hated publicity and living a *public* private life. But like it or not, we were now very public people, trying to live on as if nothing had really happened to change our lives--as if we were still private people with all the benefits of not being scrutinized--but that was wishful thinking and a time long gone, which every newspaper since November 13th, 1925, could testify to. We had never had anything before. So we were a lot like kids in a candy store for the first time. And the papers made money off of our joy. We bought Mildred a million dollar collection of the world's finest Stradivari violins when she was eleven; and Frank bought himself a fleet of Duesenberg

PASSIONS AND PREJUDICE

automobiles because he loved cars. The papers carried it all.

When the great pianist, Ignace Jan Paderewski, came to Beaumont, we hosted him. Paderewski wanted to see one of Frank's Spindletop oil "gushers," as he called them. This embarrassed Frank since there were no real gushers at Spindletop, only great producing oil wells. But Frank, who was always modest, couldn't bring himself to tell the wide-eyed Paderewski that so he arranged for a special gusher to "gush" by rigging up a four inch connection to a well-head. Frank then instructed one of his men as to what to do when he brought Paderewski out: they were to turn on the "gusher" as Frank's car rounded the corner with Paderewski. To that end, Frank had his men stationed strategically along the route for signaling the man who then, on cue, turned on the "gusher."

Paderewski was "wowed" at the sight! He even turned to Frank in his amazement at the roaring stream of spouting oil and said that that was the kind of music he liked to hear best. This gave Frank a real charge, since Paderewski was one of his favorite pianists.

Frank and I found ourselves cast as philanthropists and patrons of the arts in Beaumont. We sort of fell into the role without meaning to, and made the best of it--and were

THE 13TH OF SPINDLETOP

honored to do so. These years--the years between 1929 and 1931--were later christened "The Golden Age of the Arts" in Beaumont, Texas.

> Mr. and Mrs. Miles Frank Yount
> Miss Mildred Yount
> presenting in recital
> Paul Kochanski
> Pierre Luboshutz, accompanist
>
>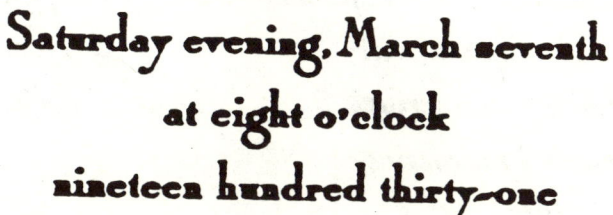
>
> Saturday evening, March seventh
> at eight o'clock
> nineteen hundred thirty-one
>
>
>
> 1376 EL OCASO Calder

PASSIONS AND PREJUDICE

Old violins from which selections will be made

	Name	Year
Joseph Guarnerius	"Del Jesu"	1741
Antonio Stradivari	"Spanish"	1689
Antonio Stradivari	"Swan"	1737
Antonio Stradivari	"Wilhelmj"	1725
Joseph Guarnerius Del Jesu	"Chrysler"	1737
Antonio Stradivari	"Reynier"	1681
Antonio Stradivari	"Piatti"	1717
Dominicus Montagnana	....	1735
J. B. Guadagnini	....	1782
Andre Guarnerius	....	1688
Joseph Guarnerius Felius Andre		1715

THE 13TH OF SPINDLETOP

--a--

Praeludium E major . . . . Bach
Sicilienne e Rigaudon . Francoeur
Rigaudon . . . . Rameau
Chanson Louis XIII e Pavane . Couperin
Praeludium and Allegro . Pugnani

--b--

Andante a la Zingaresca . Dohnanyi
Flight (Dedicated to Col. Lindbergh) Kochanski
Waltz A major . . . . Brahms
Ritual Fire Dance . M. de Falla

PASSIONS AND PREJUDICE

Through all of the arts events we were able to bring to Beaumont, Mildred was always by Frank's side. It was natural that Frank's love of the arts would rub off on her. She was plainly and simply a promising, budding artist on every front, which made me proud. She played the harp with great beauty and grace, and had already painted some extraordinary oils by age ten. She was fast becoming an auburn-haired beauty. She loved her daddy, and her daddy doted on her. She was, after all, our one and only child, more treasured than any oil.

Frank was a self-made oil man who had become an extraordinary man of culture and refinement during these years through his own self-study and desire. All his free time was spent reading and studying, and he became an authority on literature. He was a connoisseur of the arts, loved music passionately, and he was a collector who collected everything from precious artworks and paintings that ranked among Texas' finest to, as I said, a fleet of Duesenberg automobiles. His favorite car was a black Duesenberg straight eight of 265 horsepower, capable of a speed of ninety miles an hour in second gear and 135 miles an hour in high. Frank was big on speed. He also loved a Cord front-drive car that was a dark maroon--one of my favorite colors.

One of the most meaningful honors ever bestowed on Frank come in January of 1931,

THE 13TH OF SPINDLETOP

when Governor of Texas Ross Sterling appointed Frank to the Board of Regents of the University of Texas. Because of his self-education and the importance he placed on education from his background of so little formal education, this honor meant, I think, the most to Frank. He was to fill a term that would expire in January 1937. As a U. T. Regent, Frank got to serve as chairman on several key committees, including ones in charge of University lands and in charge of the Medical Branch of the University at Galveston.

Honors seemed to pour in upon Frank from every direction. He was a Rotarian and an Elk, and on June 13th, 1933 (another one of those 13ths in Frank's life), the Beaumont Rotary Club gave him the title of **Distinguished Citizen**. Frank deserved the honor. I got the consolation prize as his wife--a basket of roses.

Where I was, you can bet animals were not far away. I couldn't live without my animals, and I was fast populating our Calder Farm with all kinds of stock, many of which were entered in the South Texas State Fair, which some people found humorous. I always made sure that our Jersey cows wore light khaki blankets to protect them from bugs and other pests and to enhance the shine of their coat so they looked good at the show. People said I made them wear khaki "kimonos." And I always made sure that they were all bathed, clipped, and everyone had her hooves

PASSIONS AND PREJUDICE

manicured for the thousands of people that would be seeing them at the fair. We only had registered Jerseys, my favorite being "Louise." But the eighty-acre farm we had on Calder was by no means limited to Jerseys. We had chickens, ducks, and turkeys galore, and a few peafowls. I put a lot of worth in animals. Animals are true-blue, noble in the best sense. Their feelings never change toward you, they're honest, and you can always count on their love and loyalty. On a scale of one to ten, I'd say they rank a couple of notches above most humans.

The barn was complete, and Spindletop Stables of Beaumont was ready and waiting to receive its first guests. That's when Cape Grant, the trainer Frank had hired and given a blank check to, to find some show horses for us, pulled up into our driveway on Calder with a fancy horse van in tow and a new car that he had taken the liberty of buying with part of the blank check money Frank had given to him. Only the best for the best it seemed. I had no idea about the horses this man had in the van, but standing before me that day was a horse of a different color. Of that I was sure. It was the first time I ever set eyes on Cape Grant. It was June 8, 1933.

Cape Grant was clearly the "Rhett Butler of the times." He was a good six feet plus, as dark and handsome as they come, and equally disarming in his manner. He was a showman from the top of

THE 13TH OF SPINDLETOP

his head to the tips of his toes, with a charm, flashy smile, personality, and ego as big as Texas itself. That multifaceted mixture that went into the makeup of what was Cape Grant literally leaped at you with a presence bigger than life when he was in the show ring with a horse as magnificent as a Saddlebred. And in the show horse ring, he was a force to be reckoned with and nobody's fool. This one was a ladies' man, immensely handsome, who dressed to the hilt as a dandy, and who appeared to me to be a close cousin of a tomcat who liked to drink and carouse around all night--among other things. He was nothing short of a fascinating, colorful character that was never disappointing--a real crowd pleaser. He not only knew he was good, but liked being important. When we were somewhere together, Cape always liked to be paged over the loudspeaker over and over again--ten or twelve times was not uncommon. He plainly liked to hear his name announced in crowds to show what great demand he was in and just how important he really was.

Cape Grant knew what he wanted, was unabashed when he set his sights on it, and was going to get whatever it was come hell or high water. And one thing that he liked was *nice things.*

He was born William Capers Grant January 7, 1899, in Little Elm, Texas, outside of Dallas in Denton County. His father was a general merchant and preacher from Muddy Creek, Tennessee; his

PASSIONS AND PREJUDICE

mother Mary Alice Reynolds from Howe, Texas, had some education from a college of industrial arts in Denton and saw to it that Cape was well educated.

With a father from Tennessee, little Cape couldn't help but be acquainted with Saddlebred horses. He had been winning blue ribbons at horse shows since he was eleven years old. Most of the people he was around were cowboys, and he preferred to show with western saddle like the big cowboys; but when Cape came in last showing a horse that way, he changed immediately to English saddle and won first place at the next show. This was a lesson that left a heavy mark on Cape's life. He wanted to win, and was willing to do whatever was necessary to make sure he won. It was his overriding trait. He was a natural in every sense of the word and in a class unto himself when it came to showing horses to their best advantage. For the next three years after winning his first blue ribbon in 1910, he won top honors as the best child rider at the Texas State Fair in Dallas. He worked for several stables as a trainer and enjoyed some success in his career, but he wanted even more of the bright lights, fame, and fortune that winning could give him. He was ambitious with the talent, merit, feel, and flashiness to garnish it and shine. Beyond a doubt, I think he was the best all-around rider that ever got on a Saddlebred and entered a show ring in Saddlebred show horse history.

THE 13TH OF SPINDLETOP

Cape's disposition for the finer things in life was already in place as a boy. It was his nature. A favorite story that circulated about Cape's first coming to Dallas and his working with his father was that he didn't show up to work at the barn one morning. So the owners went searching for him. When they finally found him, he was back at the fancy hotel where he and his father had been temporarily put up in, sitting in the middle of the big bed eating chocolates like a little king.

He even dressed like a king, complete with elegant manner and presence; and he had "chutzpah." When Cape Grant showed a horse, he dressed the part complete with high silk hats and everything that went with it. To his credit, it was Cape that started proper show dress in the ring in the Southwest, giving horse shows that added edge of glitter and glamour. Cape Grant plus a beautiful American Saddlebred equaled nothing short of magnificence incarnate. There was none other like him. And there has been no other rider like him since.

For a little fun and sport, Cape used to get decked out in his finest and go to rodeos and sign up to ride broncs. With the harmless charm of an adder and a likableness that got him far, he would let the cowboys tease him mercilessly about his dress and then get them to place bets that this dandy couldn't stay on a horse, let alone ride a wild bronc. After they put up enough money, he

PASSIONS AND PREJUDICE

would step up to his turn and beat them all hands down. He knew exactly what he was doing and could out-distance them all. Cape earned a lot of pocket money this way, and he liked money and *nice things*. It was always a mistake to judge Cape Grant by his cover, and the faster a person learned this, the better off he and his pocketbook were.

Not only was Cape immensely handsome and intelligent, he was married to a drop-dead, knock-out of a brunette by the name of Nola, who could match the beauty of any magazine model. They were close in age and had married when he was twenty and she was nineteen. She was an extraordinarily nice person and good mother. And when Cape and I first met he had two of the finest little sons imaginable: Silas, the oldest, and William Capers, Jr., named after Cape. His family would be joining us in Beaumont, with Cape as our trainer and manager for the stables.

Cape Grant was twelve years my junior, and twenty years Frank's junior--so considerably younger than both of us.

Frank and I watched as Cape opened the van's doors and brought out what were to be the cornerstone Saddlebreds of Spindletop Stables. They were the epitome of nobility. First, was Beau Peavine. There was probably never a more beautiful Saddlebred horse on earth! He was an unparalleled golden chestnut stallion with three

THE 13TH OF SPINDLETOP

white socks, star, and snip, with a heavy red mane that in the right ring lighting took on a magnificent golden glow, almost as though there was some type of special heavenly aura about this horse that we were allowed to see shining through for a few brief moments when he showed.

Cape had gone all the way up to Missouri hunting for the best possible show horse prospects. And with his keen, experienced eye, he had found Beau Peavine at the Woolford Farm there. And only a short way down the road from where he found Beau Peavine, he found a three-year-old, five-gaited stallion.

Cape said the Missourians called the five-gaited stallion "Chief." When I looked into those intelligent, deep eyes and saw how he listened with those alert, sensitive ears, and when I beheld the strong grace of the muscles and veins of this great animal in relief like in a da Vinci painting, I couldn't help but murmur under my breath to him as I patted his long, majestic neck, "You'll be called Chief of Spindletop from now on, boy." I believe he understood, and was proud.

I would often look out of my bedroom window and watch Cape working Beau Peavine and Chief of Spindletop in the first light of dawn in the stable ring. Cape was always an early bird, and he preferred to work in the crisp cool of the morning when the light fog was just beginning to

PASSIONS AND PREJUDICE

clear, and otherwise everything was new and fresh and golden and beautiful--and wrapped in a peacefulness and quietness that was as beautiful and pure as a newborn sleeping. Only the snorting of the horses in the early morning air and the rhythmic pounding of their hooves on the soft ground broke the morning silence, as though composing some secret symphony so sacred that it was being stored in their great hearts for safe-keeping. The fog from the heat of their nostrils meeting the cool morning air and their powerful da Vinci lines beginning to shine with sweat, like oil that had been rubbed on the well-developed, muscular bodies of Olympians, put a crown of spectacle and worth on each day's dawning. The picture was nothing short of a romantic portrait of the passion of the world's greatest lovers, with Cape and the horses becoming more and more one with each breath and movement, in an ecstasy of perfect and eternal harmony and concentration that was breathtaking. Looking from my window, I felt privileged as though looking through some secret keyhole and beholding one of the most intimate and beautiful moments two lives could ever share together--a moment not made for other eyes.

Cape offered to teach me to ride. I wanted to learn.

At first, just getting up on a Saddlebred horse was a challenge for me. Somehow the

THE 13TH OF SPINDLETOP

stirrups seemed way too high up, and the horse--well--, it was like trying to mount the Statue of Liberty, Saddlebreds were so tall. I thought a ladder was called for. "Nope," said Cape cupping his hands like a stirrup, "just put your foot in here and hop about three times on that other foot and throw yourself over. It's easy."

So, I put my foot in his cupped hands, with another farm hand holding the reins of the horse and steadying him, hopped three times, and threw myself all the way over on top of the horse, falling short and floundering, with both Cape and the farm hand trying to push my "bee--hind" up; but it was too much for both them, the horse, and myself, and I slipped all the way past the saddle and landed on the other side of the horse on the ground, a far cry from my original destination. I was fully humiliated. The farm hand and Cape, trying to keep straight faces to cover-up the laughter inside that kept coming up in their throats like burps or oncoming sneezes someone was trying to hide, looked at each other. Then Cape, always in control, said, as I got up dusting myself off, "Well, I'm glad that's over. Now that you've fallen off for the first time, we can get on with it. I have to admit though, most people wait until they are up in the saddle and actually on the horse before they fall." The farm hand could take it no more, and turned around in an attempt to further control his "laugh burps" that were coming up

PASSIONS AND PREJUDICE

more uncontrollably, frequently, and practically folding him up.

"Go ahead and laugh," I said, giving us all polite permission for what was inevitable. I started to laugh at myself; and Cape and the farm hand looked mighty relieved to finally burst out. I think even the horse may have cracked a smile.

"That's a mount that will go down in history," Cape said, glancing at the same time at an old trunk in the corner. "Here, let's take the horse over to that trunk.----Now, Miss Pansy, you climb up on the trunk, and we'll help you on the horse."

I did, and I was up in the saddle--on the second try. I don't think the horse, Cape, or anyone else ever forgot my first keen example of equestrianship. My first time on a Saddlebred was told around the dining room table over and over again, and probably whinnied about in barns among Saddlebreds for years. Imagine Beau Peavine quipping in the barn to Chief: "Uh-oh,--there she comes again. Who wants to be the lesson horse today? (All horses turn their butts and tails to the front of their stalls, and flatulate in unison.) Does that mean no volunteers?" But I was undaunted by this disastrous start. I had a good sense of humor about myself. And I got better and better--even downright good to the point of becoming a fair competitor. I was used to getting up, dusting myself off, and getting back on by now.

THE 13TH OF SPINDLETOP

That was June. Amazingly, by September Cape said he thought the horses were ready and took them to the World's Fair in Chicago. The horses did well for a first time out with Cape, with Beau Peavine winning second in the junior five-gaited stake and Chief of Spindletop outdoing himself by taking first place in the three-year-old, five-gaited class. Not bad at all, and we could not have been more the proud owners. We were more like kids in a candy store again--overjoyed. Cape, Frank, and I talked it over and decided to ship the horses directly to Kansas City from Chicago to the American Royal in November--another world-class premier horse show for Saddlebreds. This was one that Frank eagerly awaited; and we all planned to be there and watch Cape show Beau and Chief for Spindletop Stables. It was hard to wait.

That mysterious date of the "13th," that literally haunted Frank all his life, came up again. It was the Monday before the American Royal that was going to be held in Kansas City--November 13, 1933. It was also eight years to this very day that Frank had brought in Spindletop--November 13, 1925. It is peculiar how the 13th played its ghostly strains through Frank's life. We were having a small family party that night. We were all gathered in the living room--Frank, me, Mr. Westenhoeffer, our attorney, and my brother Emmett, and Beulah, our cook, who had been with us for years, and who we affectionately called "Boo." Whenever

PASSIONS AND PREJUDICE

Frank and my brother got together it was always a hoopdala. Frank was in fine form that evening bantering with Emmett and telling one joke after another. He had us pretty much in stitches. Emmett asked him why he didn't go into thoroughbred race horses. Frank picked it right up with playful tongue in cheek without dropping a beat. "Well," he said, "we had a thoroughbred once. He did a mile and a quarter in two minutes flat.----Then we took him out of the van, and he didn't do so good." Everyone laughed. He seemed to be a well-satisfied, happy man that night and looked especially handsome dressed in his favorite white linen suit. He always went out to the oil fields to check on the operations every morning at promptly ten o'clock in a white linen suit and big Panama hat. That was his trademark among the crews. I even had an oil portrait painted of him in that striking suit and hung over our mantel. We were celebrating the upcoming American Royal in Kansas City that was coming up on Saturday; and Frank was really looking forward to that trip.

Frank asked Mildred to play her Stradivarius for him that evening. So Mildred got the violin out of its case, and she played so beautifully it seemed to touch Frank particularly deeply and brought tears to Frank's eyes. It was obvious that she was forever the love of his life. Frank had even had an apartment complex erected in Beaumont from granite imported form one of our

THE 13TH OF SPINDLETOP

own quarries in Colorado and named it "The Mildred" after his little girl. And it was just as obvious how much Mildred loved her father! With every sensitive touch of the violin strings with her agile and delicate fingers, with every strain of music that she sent out that night from that carefully crafted instrument that Frank had gotten her, there seemed to be a whispered message from her little soul to Frank's soul: and it said, "I love you, Daddy." I noticed about this time that the punch and cookies we were enjoying were running out; so I told Boo to just keep her seat, and I would bring in some more from the kitchen since I was headed that way anyway. I went into the kitchen, and suddenly Mildred ran in with tears in her eyes, grabbed me by the hands and pulled me toward the living room, screaming hysterically, "Mama! Mama! Come quick! It's Daddy!!!"

I ran into the living room. There was Frank on the floor--dead. He had had a massive heart attack!

It was as though my entire spirit and being rebelled at that moment against the realization of Frank's death with a scream from the very depths of my soul. It was not acceptable to me; and nothing nor nobody could make me accept Frank's death during those moments. I gathered a certain unreal calm about me to defend myself against the rage I felt at the suddenness and injustice of death. My surroundings were cast in an almost

PASSIONS AND PREJUDICE

surreal light. With all my spirit, I denied death's cold grip on my life and dominion now over Frank's. I called the servants and told them to take Frank's body upstairs to the bedroom. Everyone seemed frozen, stunned. I followed behind the servants carrying Frank's body up and went in and locked the bedroom door behind me; and I sat there in the dim light of the darkest night in my life with Frank's body laid out before me on our bed. I sat there all night long, as the others kept a vigil outside, not knowing what to do. I stayed there with Frank and wouldn't let anyone in until the next morning when Monsignor Kelly from St. Anthony's Church came and knocked on the bedroom door. He talked to me softly, and I opened the door slowly, reluctantly. I was cut by a depth of pain and despair I had never before known. He extended his arms, and I literally collapsed in them in a passion of tears as though they were arms extended by God himself as a last refuge of hope; and at the same time I let my rage loose through my tears for the first time, pounding on him, crying out "*Why!!!*" at the top of my lungs, as if brazen enough to pound on God himself and demand that answer that we all want, but none of us ever get. It seemed that those who were most dear to me were always taken away on a thirteenth. My mother died on a thirteenth, my father died on a thirteenth, and now Frank had died on November 13th. But why, indeed. Oddly enough, Frank had just had a complete routine physical and gotten a clean bill of health from our

THE 13TH OF SPINDLETOP

doctor only the week before his death. And why, indeed, did he die on the eighth anniversary--to the very day!--of his bringing in Spindletop? And why--why was the *13th* chosen as the bearer of so much joy and so much tragic pain in Frank's life and in my life? These *whys* remained the secret of God, and among the mysteries and secrets of Spindletop itself. I somehow believe the Hill alone was privy to the answers.

The men went in and took Frank's body away to the Pipkin and Brulin Funeral Home.

Word had already reached the papers. It was Mildred--red-eyed, grief-stricken, and hurting--who brought the paper in that morning. But there was another look on her face that morning, as she peered at the front page headlines, that startled even the depth of my pain. She looked at me, pointing to the paper in horror, her voice strained with the whisper of disbelief, "Mama, Mama--what is this!?!" There were three bold headlines that took up the front page of the Beaumont newspaper that cold morning:

PASSIONS AND PREJUDICE

MILES FRANK YOUNT DEAD AT 53

PANSY YOUNT PROSTRATE WITH GRIEF

MILDRED YOUNT ADOPTED

Mildred was thirteen. Frank and I had never told her that she was adopted. It had been a fragile secret sealed with love and kept safe deep within our hearts--we thought for her own good. No one spoke much of such things in those times. How could the reporters have been so cruel and inhuman! It was the second most difficult moment in my life, as I felt Mildred's pain on top of Frank's death. It was like the press had become judge and jury; and in the interest of all the citizens of Beaumont who wanted "the news," they had put us in front of their firing squad of headlines and had sentenced us to a slow, torturous death in the most unconscionable, cold way possible--all for selling newspapers. Mildred was hysterically crying. And suddenly a whole group of uninvited reporters pushed their way through the unlocked front doors of our home with flashing cameras and a barrage of indelicate questions. I screamed

THE 13TH OF SPINDLETOP

at them to get out of my house. They were nothing but vultures. Frank and I had never wanted publicity, and we certainly did not want it now. I threw something at them in my rage--what, I don't even remember--and pushed them out the doors, slamming the front doors behind them and locking them, and then collapsing against them in an anguish I would wish on no one.

I sat Mildred down with my arms around her and tears streaming down our faces and tried to explain to her. Yes, she was adopted. But no one could ever have loved her as much as Frank and I; and she could never have been more ours than if she had been our own flesh and blood. I told her that I supposed we didn't tell her because she was ours; and we were somehow deep down afraid that she wouldn't love us as much if she knew. Silly thoughts, I know; but we had had such thoughts. We loved her so much, and her daddy loved her more than anything in this world. We held each other in our pain; it seemed all day long.

As the newspapers had reported time and time again:

> **Mr. Yount is said to have a passion for music, and it is reported to be Mrs. Yount's ambition to see her daughter become an artist.**

PASSIONS AND PREJUDICE

After Frank's death, however, Mildred never touched the violin again, or the harp, or painted anymore. It was like all that she and Frank had shared creatively together died within her the moment he died.

An important part of me, too, died that November 13th with Frank, and died again when Mildred saw those headlines the next morning. I was left completely numb, a shell of myself. I wish it could have been me instead of Frank. Frank and I had been more than husband and wife in our eighteen years together. We had been friends, companions, dreamers on horseback.

Flags flew at half mast in Beaumont. Frank loved Beaumont, and Beaumont loved Frank. It was Frank and I that always made sure not one person in the company from the elevator boy and janitors to the top executives were ever forgotten. Each, without exception, always received a substantial bonus at Christmas time; Frank's crew was always given the best work boots that money could buy and Stetson hats in addition to bonuses. It was Frank and I that made sure during the Depression that each and every city employee in Beaumont was paid by meeting the payroll when the city of Beaumont itself couldn't; and we always made sure during the Depression that no child went without Christmas presents or proper nourishment.

THE 13TH OF SPINDLETOP

I told Cape Grant to go ahead and show the horses in Kansas City at the American Royal, just as Frank had wanted--and ride them with a vengeance to win, dammit! I immediately set about building for Frank the finest mausoleum money could buy. I wanted it to last forever.

On the day of Frank's funeral, it was partly cloudy over Beaumont, with a light southeasterly wind across the coastal area. Over twenty-thousand people gathered on the lawn outside our home and lined the streets of Beaumont for Frank's funeral. It was as though Beaumont was spilling over at the seams with grief. There were loud speakers set up so all of the people filling the yard and the streets of Beaumont could hear the minister as he conducted Frank's service.

At the very stroke of three that day, as the service for Frank began in Beaumont, at the very moment Frank would have been in Kansas City himself watching the horses show, and at the very moment all the people in Beaumont were crowding in, straining to hear the minister's words, Cape Grant was riding into the ring at the American Royal in Kansas City with an exhibition of horses that would go down in Saddlebred history. Those horses won every conceivable first place that day, just as if they knew, just as if they understood, that they were building their own precious monument to Frank in a eulogy of eternal beauty and grace before thousands of spectators that

PASSIONS AND PREJUDICE

were holding their breath with heightened excitement at the tragic circumstances under which the horses were performing. It was a monument that would have made Frank proud!

When Frank died, I was determined to follow through with building a mausoleum to house Frank's body, as a monument in Beaumont to his memory. I talked those in charge of Magnolia Cemetery into developing a piece of land across the street from the actual cemetery, across Pine Street, where the mausoleum would stand alone with nothing in the front of it or to the sides of it or across from it. I paid for the development, and I hired workmen to build the mausoleum with a chapel that would have bronze doors and stained-glass windows. There would be a breezeway and a family mausoleum to the side. The mausoleum would be made of granite, and the interior would be finished in Colorado Greenstone imported from our own quarries in Manitou, Colorado.

When the mausoleum was finished, I had Frank's body transferred from its temporary site in Magnolia Cemetery and reinterred in the mausoleum, with clergy present.

There was a hurricane that hit Beaumont about this time, and I went out afterward and

THE 13TH OF SPINDLETOP

inspected the mausoleum to make sure everything was in order. I noticed something on the floor inside. It was sand. I asked the workmen about it; and they said that with the hurricane and all, a little sand had come down from the ceiling and landed on the floor in places. I asked them point blank how long this mausoleum was supposed to last, and they said, "Why, Mrs. Yount, six or seven hundred years."

"Six or seven hundred years! That's not long enough. Tear it down," I ordered. They argued that six or seven hundred years was a long time, and it was Depression times, and all the money that had been spent. "I said tear it down, dammit," I screamed. "It's my money! I don't want anything with the Yount name on it falling down when I'm gone after only six or seven hundred years. Tear the damn thing down!!" It had cost around one hundred thousand dollars to build, and it was 1934, which meant about sixteen million dollars in today's money. But no matter; in my eyes, it was not good enough for Frank.

I started over. This time, the mausoleum, with gold and copper inlay, would be built in the ground to last forever. Meanwhile, I moved Frank's body temporarily, again with clergy present, into Mr. Westenhoeffer's mausoleum that was not in use at the time since Westenhoeffer was still living; then Mr. Westenhoeffer died, and I moved Frank's body again, this time temporarily back into

PASSIONS AND PREJUDICE

the old Magnolia Cemetery, and again with clergy present.

All of Beaumont was talking, "Crazy Pansy. She's already moved her husband four times. She can't seem to bury him." But the day came, and the new mausoleum was at last finished, and I did lay Frank to rest, again with full religious services being conducted by the clergy. It was different than everybody was saying, though. Somehow, I think I thought, somewhere deep within the recesses of my mind, all neatly rationalized with a bow on top to make it all right--that as long as I kept building--somehow--I wouldn't have to let Frank go. I never accepted his death and that it was beyond my power to keep him here on earth with me through my own efforts. I was not God. It was time for me to fully realize that. And the day did come--and--I did let Frank go. It was 1934.

Although in excellent health, and although his early death came as a complete shock to us and everyone else, Frank had made out a will the year before he died. In it, he had seen to it that Mildred and I were well taken care of.

Frank's eighty-four year old mother, Hattie, put the icing on the cake. She was the only one to try to sue me over the will and the money. I guess you could say, she never liked me none too much.

THE 13TH OF SPINDLETOP

Where there's money, there's vultures. Some look the part, some don't--it's hard to tell who's who. It seems to go with the territory though. Headlines, such as the one in the *San Antonio Express* newspaper on April 13, 1934, which were carried in all major city newspapers nationwide, only rush these vultures to the carcasses a little faster, with all the accompanying consequences for the victims. At times like these it's good to remember the old adage: "*It looks like a sheep, it walks like a sheep, it sounds like a sheep--but it's not a sheep! It's a wolf!*"

INVENTORY OF FATHER'S ESTATE SHOWS THAT 13-YEAR OLD MILDRED YOUNT TO BE WEALTHIEST GIRL IN TEXAS

As a result of such headlines, marriage proposals poured in from people Mildred and I had never heard of, let alone set eyes on. For a thirteen year old, they were ridiculous, and no less so for me. It was more than obvious that what everyone was proposing to was my money. And my money said "no."

The life Mildred and I were leading now was a little drab and like being in a fish bowl for a thirteen-year old girl. After Frank's death, I had no

PASSIONS AND PREJUDICE

choice but to take Mildred out of public schools and to get private tutors to come to the house for her own safety. This was the time when the Lindbergh baby had just been kidnapped, and I feared someone might just try to kidnap Mildred for ransom. There were crank calls and death threats from deranged minds out there, all of which were communicated to the police. And although we had German Shepherd guard dogs at the house and an intricate burglar alarm system for protection, bodyguards were a must everywhere Mildred went. Mildred would often sigh and say, "It doesn't seem as though I'm like other little girls. I can't even play without being watched, and now I can't even go to school with my playmates." So eventually, I sent Mildred to one of the best and safest private schools in the country, the Hockaday School in Dallas, and paid the way for her best little friend to go through the school with her so she would feel more comfortable with being away from home. I missed her, but felt it was in her best interest. She had to have a life of her own.

There was nothing really left for me in Beaumont but memories. Mildred was in the Hockaday School in Dallas; Spindletop horses had won everything there was to win in Texas and Louisiana and were making a name for Spindletop Stables around the country. I was the first woman appointed to the Board of Directors of a major United States bank, the First National Bank of

THE 13TH OF SPINDLETOP

Beaumont; and I was signing Spindletop oil leases to the giants that it birthed, such as Gulf, Texaco, and Sun Oil. I was getting richer by the minute. But these things were somehow empty to me, now that Frank was gone and Mildred was in Dallas.

Frank and I had talked a lot about Kentucky and Saddlebreds when he was living--what we wanted for Spindletop Stables. He had always said that Kentucky was the place to be if you wanted to make your mark in the Saddlebred world. We had already made Beaumont a new stronghold and American Saddlebred center for the country, a real feather in the Texas Stetson. And Cape Grant was aware that I was thinking more and more about going to Kentucky. I had already been looking for land there--a place to move Spindletop Stables. These things had been very much on my mind lately.

In August of 1934, Cape Grant and Nola were blessed with their third child, who they named Miles Frank Grant after Frank. I was very touched by this. He was a beautiful baby.

But shortly thereafter on September 1, 1934, when little Miles Frank Grant was only two months old, Cape told Nola that he had no intention of living with her anymore, and that she should get a divorce.

PASSIONS AND PREJUDICE

I was closing the second chapter in my life. I now turned my eyes and attentions to Kentucky.

PART III

Blue Blood

BLUE BLOOD

There is a poem that circulates in Kentucky that goes somewhat like this, and I never did know who wrote it, although I rather think it was someone born and bred in Kentucky:

If you're a good little boy or girl,
And say your prayers every night,
Then when you die, if you're lucky,
You might end up in Kentucky.

You have not lived or experienced excitement, or challenge, or nature's best shot at beauty and glamour, or know what life is really all about until you have lived in Kentucky. This is gospel.

Kentucky is a school unto itself. Make no mistake, there is absolutely nothing smarter than a Kentucky farmer. No one can add up what you owe him quite as fast, even if he's never set foot in school. As one farmer told me, "I lay awake at night thinkin' how I can turn a dollar from the next guy." Anywhere else you've been is a child's play pen until you've graduated from the Kentucky school of life.

I loved Kentucky and its beautiful rolling hills, white board horse fences, and lush, cool surroundings from the word go, especially coming from Texas where it was so flat you could see for miles, and scorched, burnt brown under a blazing sun was nature's fashion choice pretty much year 'round. But being perfectly frank, there was a time there when I looked at this little poem, and thought I had taken a wrong turn somewhere and landed in that other place.

You know, even the full moon looked different in Kentucky. I knew it was the same moon I saw over Texas; but whereas the one over Texas seemed so brilliantly white, the full moon over Kentucky took on the most mellow golden glow like a big, fat pumpkin sitting on the horizon when it rose. It always held me in awe.

Suddenly, it dawned on me what Frank meant that time at the Southeast Texas Fair the night I first saw a Saddlebred horse and asked him if Kentucky really had "blue" grass. I remember how he smiled at me and said, "only in the morning dew." You hear of Kentucky bluegrass, but that's deceptive if you go around actually looking for "blue" grass. What they're talking about is a particular type of lush, deep green grass, *poa pratensis* for the scientifically minded, that gets an

BLUE BLOOD

almost bluish tint to it when the dew sits on it in the morning. That's Kentucky bluegrass, and Thoroughbreds and Saddlebreds alike thrive on it as much as the grass thrives on Kentucky limestone. Thus, Kentucky's nickname, the "Bluegrass State." But if a stranger asks a Kentuckian about "bluegrass," the chances are he's not going to fool around with all sorts of explanations and scientific mumbo-jumbo, but just say, "Yep, we have it. You'll see it sooner or later; just keep lookin'." Then, if you look closely you'll probably catch a mischievous smirk on the native's face as he walks away. Since I was green and uninitiated, the natives had some fun with me, too. They told me that there were no termites in Kentucky and no snakes, either. I thought, what kind of place could this be with "blue" grass, no termites, and no snakes. I soon found out.

Now the Kentucky of my times was a very old, closed society where "real society" was used to being rich and whose money was so old that it had moss growing on it. In Kentucky society, you didn't just walk up to someone and say "Hi, neighbor!" and start telling your life's story, but you had to be properly introduced by one of the "inner circle," so to speak. Texans grinned ear to ear--meant you were friendly; but in Kentucky that could be misunderstood: it seemed that if you smiled too much, it meant you were up to something, criminal-like. Being from Texas, I had never met a stranger and was a lot like a big puppy

PASSIONS AND PREJUDICE

climbing into laps and licking faces, I was so happy to be in this place called Kentucky. My mistake was, in part at least, not realizing that I was no longer in Texas, and taking it as though I were at home in familiar surroundings. It never struck me that in going across a few state lines, I was in a different country and culture altogether, and that the rules had changed. Both Texas and Kentucky had trees and grass, but they were not the same place by far.

In hindsight, I can see that to Kentuckians, I entered the scene a little like an elephant in a porcelain shop. But I was just me. I had come from a very simple, hardworking, and poor background, and Spindletop money had not changed those facts, no matter how I dressed or how hard I might try, which wasn't much, I admit. I took people as what they were, like was the custom in the all-accepting, absorbing frontier of Texas where rugged individualism and new ideas were admired, encouraged, and celebrated. There was something downright "down home" about me; I looked different, smiled too much, I was outspoken, my ways were different, and I talked down-home, Texas lingo--it may as well have been Greek. Both the Kentuckians and I used the same words, but we weren't communicating. I think what we needed was a skilled translator that could translate Texan into Kentuckian and Kentuckian into Texan. But even that probably wouldn't have helped either: my money just didn't have moss on

BLUE BLOOD

it, but had that "store bought," brand-spanking-new smell--a little like the aromatic difference between a hand-rolled Havana cigar and a can of snuff to the connoisseur noses of Kentucky blue bloods and Kentucky society.

I was a little slow in noticing the change of social rules. You know an alligator just sits in the water with its eyes showing, perfectly still, but it can suddenly come up from behind you with its tail when you least expect it and knock you for a loop if you don't know the way an alligator operates. That's about how I had to learn about Kentucky. I started out on the wrong foot, and it took a couple of loops to get through to me that I was not in step.

The first thing I did right off the bat was to buy eight hundred acres of land out on Iron Works Pike, just outside of Lexington, the W. C. Coe "Shoshone Horse Farm." And because I didn't like just eight hundred, I bought an additional two hundred acres from neighboring farms and tacked it on to make a thousand plus acres--a thousand sixty-six to be exact. It was 1935, Depression times, and I paid cash.

Secondly, although we populated Spindletop Farms with a lot of different types of livestock, I had a good-sized herd of Texas Longhorns brought up from Texas to make the place look a little more homey. That's when I received an

PASSIONS AND PREJUDICE

anonymous note pinned to the front entrance gate of Spindletop Farms. It said:

> *Kindly remove those "things" with the ten-foot horns from the front pasture along the main road. They're scaring our horses.*

I think this tacky commentary must have been written by one of the Thoroughbred people along Iron Works Pike. I'd never know.

Then I ran up the Texas flag for all to see at the site of where I planned to build a palatial mansion that I would call Spindletop Hall, and where I would locate the new Spindletop Stables. I always ran up the Texas flag outside of our homes, because it made me feel more at home. There was one flying outside of our summer home in Rockledge in Manitou, outside of Colorado Springs, Colorado.

ESTATE NEAR MANITOU ALWAYS FLIES TEXAS LONE STAR FLAG

To paraphrase Rupert Brooke: There is some corner of a foreign field that is ever

BLUE BLOOD

Texas, for Mrs. Yount floats a Texas flag about her estate. Here, the latchstring is ever on the outside and numbered on their guest list are others far less favored with worldly goods than she.

No one seemed to mind in Colorado. If anything, they all got a kick out of it. After all, Texas is not a place so much as a state of mind. Personally, I never left home without it. On the contrary, I took it with me wherever I went. Where I was, there would always be a piece of Texas. Made perfect sense to me.

In Kentucky, however, I believe the people took all these things personally. They were exactly as patriotic about their state as I was about mine, and it never occurred to me that raising the Texas flag in Kentucky would step on social toes and be the blunt of mirage-like gossip that would dry up when I came near and give rise to secret whispers that would be carried on wherever I was. In hindsight, I guess what I would have said to any of them that got offended, had I have known, was: "Get a life, people." After all, what business was it of theirs?

Before coming to Kentucky in 1935, I had rid myself of much of my business responsibilities in Beaumont by selling my oil interests in the Yount-

PASSIONS AND PREJUDICE

Lee Oil Company to Stanolind Oil Company, whose general offices were in Tulsa, Oklahoma, a subsidiary of the Standard Oil Company of Indiana, for around $50,000,000 (about eight billion in today's market). It was a deal that newspaper headlines said literally rocked the foundations of the financial world. Therefore, I had pocket money when I came to Kentucky that made the horse people look like they were on thrift budgets. And they were not used to being shown up. That's not the way I meant it, but that is the way they took it, and there was nothing I could do about it. Suddenly they all resembled *bluegrass*-- when the dew fell on them in the morning as they cast glances at what I was up to on Iron Works Pike, they took on a very unbecoming bluish-green tint that could only be one of two things: envy or jealousy. I believe those who were a little more bluish were envious; and those who were a little more greenish were jealous. It was hard to tell. But I do know that not one of the Kentucky blue bloods or horse people looked good sporting those silks.

    I poured money into the building of Spindletop Farm and Spindletop Stables outside of Lexington to the tune of bottomless wealth. I meant to make it the showplace of Kentucky and a Saddlebred operation that Kentucky and I could be proud of, just as Frank and I had dreamed together. I took up where Frank and I had left off, with a determination that had a vengeance about it. It was my passion, just as the horses were. I was so into realizing this dream that I did not even notice how the people in Lexington were looking at me. To that, I was oblivious. I honestly and naively thought they would love Spindletop and enjoy it with me. I should have remembered George Bernard Shaw's comment by Charles to Saint Joan in his *Saint Joan*, "Do you really expect people to love you for showing them up?" I didn't make the connection.

I had built three other "Spindletops" before in the form of summer homes in DeLand, Florida, in New Orleans, Louisiana, and in Manitou, Colorado, outside of Colorado Springs. Therefore, I was no stranger to drawing up plans and having my own ideas brought to fruition. I had both the drive and the push. So, in Kentucky, I drew up my plans for Spindletop Hall outside of Lexington and hired the best architect Kentucky had to offer, a Mr. Hutchings out of Louisville, and a contractor, N. L. Ross that I already knew from Colorado who had done work for me at Rockledge. The times were not favorable to women; but I was ahead of my times as a woman, I think, experienced and skilled in business by now, and determined to get exactly what I wanted done exactly as I wanted it done, which also made me less popular in a stodgy, older society like Kentucky where change was resisted with vehemence. Men, in general, though, didn't like taking orders from a woman no matter where you

PASSIONS AND PREJUDICE

were--Kentucky or Texas or anywhere else; and no matter how nice I put things, they were always less acceptable than if a man had said the same thing, point blank blunt. So I finally got wise and came to the conclusion, *"Why bother?"* Say what you want to say and let the chips fall where they fall. Chances are it won't make a hill of beans of difference in the end anyhow. I was hardly a front-runner in a popularity contest, I know. And there were times when I know, too, that both Hutchings and Ross thought I was intolerable; but the way I looked at it, we were going to get this thing done right--and the way I wanted it--come hell or high water.

This was Depression times. And when I first met Mr. Hutchings in Louisville, he was wondering where his next building project and dollar were coming from. No one was building. No one had the money. I sat in his office one afternoon, having made a 2:00 p.m. appointment. It became 2:30, then 3:00 o'clock, then a quarter to four. I watched one man after another being called in while I just sat there. After more than a reasonable wait, I went to Hutchings' office door myself and knocked.

"Mr. Hutchings," I said. "I'm Pansy Yount of Spindletop. And I had a 2:00 p.m. appointment with you."

BLUE BLOOD

Without even looking up, he waved his hand at me as if to dismiss me; his other hand stayed on his head like he had a really big headache. "Go, home, Madam, and tell your husband to come in."

"Mr. Hutchings," I said. "I am a widow. And I think we have a problem."

"What problem, madam?"

"That I am a woman.----Mr. Hutchings, I may be a woman, but I am a very capable woman. I was the first woman to serve on the Board of Directors of a major United States Bank--the first National Bank of Beaumont.----Of course, I owned it.----And Mr. Hutchings, my horses have won everything there is to win in Texas, Louisiana, and Chicago--and they're going to win it all here, too. And Mr. Hutchings, if you look at the books on oil leases in the court house in Beaumont, you'll see my name by leases to Gulf, Texaco, Mobil--I made those companies what they are today--Spindletop Oil." I then rolled out my plans before him. He sat a little more erect. "Now these are the plans that I have myself drawn up for this house I want to be built outside of Lexington. Spindletop Hall."

"Built? Your plans? You want to build something, madam?"

"That's right. Build."

PASSIONS AND PREJUDICE

"Build, madam. We'll, now--build, that's what we do, all right." Hutchings jumped up bright-eyed, now smiling. His headache suddenly seemed to take a turn for the better. He took his jacket from the coat rack and put it on, buttoning it proudly, as he simultaneously pulled me up a chair.

"Have a seat, Mrs.--uh--"

"Yount, Pansy Yount."

"Yes, Mrs. Yount." Clearing his throat, "Now, about how big do you want this house to be?"

"Oh--about three large kitchens or so, a music room, library, a living room out of hand-carved oak, I'd like. Say a formal dining room, powder rooms, around seven bedrooms, three four-room suites, maybe a lounge on the lower level decorated in the colors of Spindletop Stables--blue and red, which are also Texas flag colors, as you know--haven't found a way yet of incorporating the Texas lone star, though; maybe you can make a suggestion there; uh, a bar, dog room for guests to park their dogs in when they visit, wonderful ballroom for parties--with a 'give' in the floor, so people's feet won't get tired when they dance. Can you put a 'give' in the dance floor, Mr. Hutchings?"

"Uh, yes, I suppose we could use some cork under it. That should do it, I think. Something like that."

"Good. Servants' quarters, too. Somewhere along these lines." I noticed Hutchings gulping as he looked at me. I took the liberty of tacking down the plans a little flatter on his desk with some books nearby. "Now we can have a look-see at these plans. You can see more than just talking about it." Pointing a few things out, "Here you see my daughter's suite at the top, the winding staircase that leads up from the foyer here." His eyes got very large, and he seemed nervously excited. "When could you start?"

"You want me to build this?"

"Yes. You have a good reputation in these parts, it seems."

"Well, uh--not much being built during this Depression. Hit us right hard here in Kentucky.---- Uh, start--uh, well--How's your credit?"

"I'll pay cash if that's all right with you."

"Cash?--All right?" Now he seemed almost giddy. "Uh--. You'll pay cash? Yes! Very all right, madam."

"So, when can you start?"

PASSIONS AND PREJUDICE

"I'd say,--uh--tomorrow. How's that?" We shook hands, and it was done.

Hutchings and I met with Ross, the contractor, the very next day and they got to work right away on the plans I had painstakingly drawn up. As for me, I was out there every day overlooking the construction of Spindletop Hall and Spindletop Farms--first-hand like.

I opened an account at the First National Bank of Lexington. I drove up one day and went into the posh bank. I had my cowboy boots on and may have been a little dusty since I had just come from the farm where the men were working on Spindletop Hall. I do recall everyone turning around and looking, but I didn't pay them any attention. I walked straight up to a young teller behind the counter and told the young man I wanted to open an account. He looked at me head to toe, and said that to open an account I had to have at least fifty dollars. I told him that was not a problem. He then proceeded to tell me that I would have to show him my fifty dollars before he could open an account for me. He stuck out his hand. I handed him a check from my bank, the Chase Manhattan Bank of New York, that I had folded up and put in my purse. He unfolded the check, looked at me wide-eyed, and then said, "Wait just one moment." He then went out a door behind him, and seemed to take an unusual amount of time. Meanwhile, I chatted with the other people in nearby lines. "Slow, ain't they?" I said. "They must not do too many accounts." For the most part, they smiled meekly and turned the other way as if busy with some imaginary playmates.

Just then the young man came back through the door, and said there was someone he wanted me to meet. I asked who; he said the President of the bank. I told him I would be mighty pleased to do that.

I later found out from my banker at Chase Manhattan why the young teller had left me standing so long. The young man had gone back to the back office and called the Chase Manhattan Bank in New York and said there was this dusty woman in cowboy boots at the window who was not from Lexington who wanted to open an account, and had given him--and he was laughing when he said this to my New York banker, the banker told me--a check for $30,000,000 for openers. My Chase Manhattan banker was amused at the call and told the young teller that that sounded to him like Miss Pansy Yount; and yes, the check was good, and he could cash anything for me up to $60,000,000 on that account, and that was just my checking account. That explained it. No wonder the bank President took my hand between both of his as if I were some long lost cousin or something.

PASSIONS AND PREJUDICE

Then there was this fur shop in Lexington. I had a weakness for pretty things. I went in and looked through the man's furs, and said I would like this one--picking one that particularly appealed to me. I gave him my address and asked him to send it out as soon as he could. I later found out the man turned to the gentleman beside him who I had met once before, rolled his eyes, and said, "What, that old bag? Why, she couldn't possibly afford that fur!" That's when the man next to him set him straight, "If I were you, I'd do as the lady says. You see, that old bag could not only afford that fur, but all your furs, and your store to boot. You see, that's Pansy Yount of Spindletop."

I stopped by a china shop in downtown Lexington. It had some wonderfully beautiful china in the window. And I liked it very much. I went in, and the salesclerk was buzzing around a couple of what I guess you would call Lexington society or horse people by their dress, manner, and attitude. I waited, and waited. I wondered if I was wind or something, since no one turned to greet me or to wait on me. Finally, there was a pause, and I asked the saleslady how much the china in the window was. She looked at me from top to bottom down her well-sculptured nose, along with the other two, and then snapped, "You can't afford it." She then turned back to the other two women as though I had suddenly gone "poof," and they all giggled. I stood there for a moment, then explained in the nicest way possible that I really would like that china, how much was it. The salesclerk finally stomped over to the window with a snort a little like an old sow that had been disturbed by a gnat, and turned over the price tag. "There," she snorkeled, "Satisfied, are we?"

"Yes, 'we' are," I replied. "'We'll' take it." Then "we" went over to the counter and I took my purse--since I was tired of being more than one person--and shook it upside down onto the counter, and a roll of dollar bills fell out that I always carried rolled up with a rubber band around them to keep them together. Her eyes widened. I paid her cash as the other two women looked on, and told her to please wrap the china. Then, before leaving, I took out a couple of more dollar bills from the rubber band and handed them to her: "Here, take these, and buy you some lessons in how to treat people." And handing the same to the other two ladies who stood in silence with gaping mouths open, I said, "Here. Ya'll buy you some, too." And I left with my china.

Somehow, Kentucky and I were not connecting. I thought they were rude and trying to freeze me out with jealous, cold shoulders, and they thought I was the standoffish new kid on the block with all the new money who was crashing their long-standing social party without an invitation and trying to show them up. The truth of the matter was, I don't think either one of us knew how to talk to the other in terms the other

PASSIONS AND PREJUDICE

could understand. I full well believe in hindsight that if Kentucky and I had gotten to know each other deep down and the people had known how lonely I was and how little my money meant to me and how much their friendship could have meant to me at that time, a lot of things that happened would not have happened. But then, maybe that's just wishful thinking. I really only know that I very much wanted their friendship. I would like to believe they wanted to give it--but we both just didn't know how. This is the tragic stuff misunderstandings are made of.

I tried to join the people of the community on several occasions and do what I could, trying to make the best of things. This article appeared in the *Lexington-Herald* newspaper at one time:

NEWBORNS BROUGHT THE MILK OF LIFE

This is a Christmas story. It's about a newborn baby, hard times, and a wealthy bearer of gifts.
It was Christmas time, late 1930's. The world was just recovering from the Great Depression and about to embark upon another Great War. Few people were able to be bountiful givers.
But one woman of means found a way to give the most precious gift of all--survival.

BLUE BLOOD

Milk of Life

When Mrs. Miles Frank Yount of Spindletop Farm heard that many premature babies in Lexington hospitals needed supplementary nourishment to survive, she sent her farm manager throughout the country to purchase for her the best selection of milking goats. (Because of the curds in goats' milk, it is more digestive than cows' milk.)

Twenty Nubian goats were assembled and provided a continuous supply of fresh milk. Mrs. Yount was insistent upon having a pure product, so she set about having the first goat barn in the area constructed, based on plans sent from the University of New Hampshire. She even went so far as to have electrified screen doors installed to repel insects.

Babies Thrive

The goats loved the pasture of the famous Maury Silt Loam of Spindletop, and the babies thrived on the fresh milk that was taken into town every morning.

Many of the babies that year were able to go home for Christmas.

But any good that I may have wanted to contribute often took second seat to vicious

PASSIONS AND PREJUDICE

rumor and fabricated stories authored by jealousy itself. In Texas it had been rumored that I had been a "madam" or one of the girls who had wandered up from Crockett Street in Beaumont where all the brothels were; now in Kentucky, it was rumored that I was an uneducated, old-fat-hog-of-a-cook totally devoid of manners. The truth of the matter was that I was pretty much always a homemaker and liked it that way, and, as a homemaker I had to be a pretty astute businesswoman and jack-of-all-trades, in addition. (And by the way, just for the record, I had over 6,000 books in my personal library and loved to read, thank you. I even got a commendation from the Smithsonian Institute for my collection of Sherlock Holmes books.) It's somehow jealousy's nature to try to overpower the good with bad and think of it as more delectable, and, of course, more elevating and prestigious for the one spreading the vicious rumors.

It had been two years since Cape told Nola of his intentions to leave and instructed her to get a divorce. He had come to Kentucky with me as our trainer and manager. As someone with a keen awareness once commented about Cape, "He liked nice things." He liked them too much, I'd say. Nola waited, but after two years she filed for divorce from Cape on February 21, 1936, from Beaumont.

BLUE BLOOD

She was thirty-six, and Cape was thirty-seven. I remember because I was forty-nine that very day, February 21.

Nola GRANT	IN THE DISTRICT COURT OF THE STATE OF TEXAS
vs.	COUNTY OF JEFFERSON,
W. C. GRANT	JEFFERSON COUNTY, TEXAS

TO THE HONORABLE JUDGE OF SAID COURT:

Now comes Nola Grant, who resides in Jefferson County, Texas, hereinafter called plaintiff, complaining of W.C. Grant, who resides in Jefferson County, Texas, hereinafter called defendant, and praying for a divorce for grounds shows the Court the following:

PASSIONS AND PREJUDICE

1.

That plaintiff is now and has been a bona fide inhabitant of the State of Texas for more than a year next preceding the filing of this suit, and is now and has resided in the County of Jefferson where this suit is filed for a period of more than six months next preceding the filing of this suit, and that defendant resides in Jefferson County, State of Texas.

2.

That plaintiff would show to the Court that on or about March 5, 1919, plaintiff and defendant were duly and legally married and lived together as husband and wife until on or about September 1, 1934, when, by reason of the cruel and harsh treatment of defendant towards plaintiff, they separated and since which time they have not lived together as husband and wife.

3.

Plaintiff would show to the Court that on or about September 1, 1934, defendant became unmindful of his marital vows and began to neglect plaintiff and seemed not to care for plaintiff or plaintiff's companionship; that he would leave home and stay away from home for long periods of time and while in Beaumont would fail to come home many nights and other nights would stay out until very late; all of which had a tendency to cause plaintiff great mental distress and worry and her uneasiness for defendant's welfare caused her to suffer mental agony and affected her physical condition, causing her to lose weight, become nervous and caused a general impairment of her health. That some time thereafter, defendant advised this plaintiff that they were not making any headway as husband and wife and requested her to procure a divorce, stating that he never intended to live with her again.

4.

That defendant's conduct towards plaintiff was of such a nature as to render their further living together as husband and wife unsupportable.

5.

That during the marriage of plaintiff and defendant, there were born to them three children, all boys, namely Silas Grant, age fourteen years, W.C. Grant, Jr., age nine years, and Miles Frank Grant, age about two years; all of said children being residents of Jefferson County, Texas, and at the present time residing with plaintiff.

WHEREFORE, plaintiff prays the Court that defendant be cited to appear and answer *herein and that upon final hearing she have judgment for divorce, and for such other and further relief, special and general, in law and in equity, to which she may be justly entitled.*

Attorney for Plaintiff.

Nola was a fine woman with three of the finest sons you'd find anywhere. She deserved better; they all did. Nola's divorce from Cape was granted April 18, 1936.

Spindletop Stables were now finished, and Cape had long been in Kentucky working the horses. Nola stayed in Texas and was granted custody of her sons. Deep down, Cape seemed to be affected by the formal divorce. But you would have to have known him rather well to have seen that. His preference for *nice things*, and his stellar rise to fame and fortune nationally as a Saddlebred trainer and rider--all of which he got more quickly by ingratiating himself with me and being around Spindletop money and horses--seemed to win out hands down over whatever higher self there may have been in him. The internal struggle between what seemed to be his compulsive, blinding, and overpowering ambition and those other things in life that money couldn't buy--all of which he kept well hidden beneath the veil of his affability, showmanship, and public

BLUE BLOOD

success--inevitably pushed themselves up through the cracks like ghostly vapors in various guises and in the most dangerous of ways: three weeks after his formal divorce from Nola was made final, on May 7, 1936, Cape was drinking heavily and seriously injured himself in a head-on wreck in a car he was driving in his usual too-fast manner on his way back to Spindletop Farm from Lexington. The car went out of control, just as Cape was out of control. What is it they say? You're only as sick as the secrets you keep. A secret was gnawing at Cape like a deadly cancer that was beginning to peek its distorted, tumorous head out from under an otherwise radiantly enviable and desirable-looking public complexion. But I didn't see it then.

Cape was placed in the hospital in critical condition.

Charlie Dunn and his Dixiana Farm Saddlebreds were always in the dead heat of friendly competition with Cape and the horses at Spindletop. As a matter of fact, I made it a point at one time because of the competition between Dixiana and us to buy up two hundred more acres from an adjacent farm so Spindletop Farm would be bigger than Dixiana, at least officially on the books. Before the accident, Cape had been priming Chief of Spindletop and Beau Peavine for the World Championship at the Kentucky State Fair. It was to be a showdown between Charlie

PASSIONS AND PREJUDICE

and Cape, and between Dixiana Farm and Spindletop Farms. All eyes were on this one. Charlie visited Cape in the hospital, and joked with him that this was not the way he wanted to win the World Championship, by default. And no other horses or riders even came close to Charlie and his horses. He was right; it was as good as won for Dixiana and him, with Cape and Spindletop horses out. Charlie said he had hoped it would have been more of a challenge.

But Cape, near death according to the doctors, who didn't think he would survive this one, looked at Charlie through the pain of swollen eyes and a concussion, broken shoulder, cracked vertebrae in his back, cracked ribs, and mumbled through swollen lips, "You haven't won yet, Charlie," and attempted a weak smile. Charlie looked at Cape and told him in a type of good-natured friendliness and respect meant to encourage--that strange anomaly that sometimes exists between arch rivals when the chips are down--that he would make him a wager right then and there: if Cape recovered and won the World's Championship in Louisville, he would personally hire a jug band and serenade him on the spot in honor of the win. This was a type of fun tradition paid to winners at the Kentucky State Fair. Most of the time, members of the jug band were full of excitement and enthusiasm and low--very low--on musical aptitude. Charlie's last words under his

BLUE BLOOD

breath as he left Cape's hospital room that day were, "Poor Cape."

That wager probably saved Cape's life. There was nothing that could rally him as fast as a good bet and the sweet taste of what was so near and dear to his heart--a challenging win. Cape recovered, much to the amazement of his doctors. And when the day of the World's Championship came in Louisville, Cape was there and had both Beau Peavine and Chief of Spindletop with him-- both in top form. Cape was one man you just didn't ever want to underestimate, or it would surely come back to haunt you. And never ever say, "Poor Cape." Those could be famous last words.

Before Cape entered the show ring that night on Chief of Spindletop, he turned to me and grinned, "Watch me; I'm gonna burn 'em up." That phrase from Cape was a little like a bugle being trumpeted for a cavalry charge against a lonely Indian who had forgotten to bring his bow and arrows. He meant business. He then mounted Chief, rode into the ring, and flashed one of his well-known winning smiles at the audience. Everyone rose to their feet with a cheer and an ovation at the thrill of seeing Cape and Chief again and with anticipation of the two together showing what perfection was all about that night. The contest between Dixiana Farm's entry Night Flower with Charlie Dunn showing, and Spindletop Farms'

PASSIONS AND PREJUDICE

Chief of Spindletop with Cape Grant showing for the 1936 five-gaited World's Championship was one of the all-time, most exciting matches in the history of the Kentucky State Fair and the world of American Saddlebred show horse competitions. The two were in a dead heat that could have given onlookers a Florida suntan even in the dark. It was a fight to the finish between giants of the times under those lights, with all the glitter, glamour, precision, sweat, and controlled, determined nerves imaginable, as the audience held their breath and sat literally on the edge of their seats. Chief and Cape were spectacularly beautiful and magnificent in every move, not just man and horse, but more like one well-oiled, precision instrument straight from realms beyond, which embodied eternal grace in every muscle and line. No one could touch Cape Grant and Spindletop horses in the ring. The crowd roared.

Chief of Spindletop with Cape Grant, up, won the Five-gaited World's Grand Championship that night, and Charlie Dunn with Night Flower were reserve. Cape, with a piece of straw confidently placed between his teeth, and a broad grin on his face, got his jug band serenade from Charlie, by the way, as Cape sat down in his best show dress elegance, crossed his hands behind his head, crossed his legs in kingly fashion, and reared back in a relaxed position on some stacks of straw outside of Chief of Spindletop's stall that night. No one relished that moment and triumph like Cape.

Yes, Cape survived the wreck and made an incredible comeback. He was once again the champion, the winner. Ironically, you could say the drinking that had caused his wreck was trying to pass itself off now as Cape's savior, even taking credit for his being alive. As one very stupid nurse, who was totally in the dark about how the wreck had happened, carelessly remarked to Cape: "Well, Mr. Grant, if you had been drunk, you wouldn't have been hurt so bad. People who are drunk are so relaxed they don't get hurt as bad as somebody sober like yourself." Cape didn't correct her, but he liked to smirk about how he had put one over on her. He must have repeated the nurse's line to me at least a hundred times. But the fake savior in a bottle that he was laughing and smirking about had a flip side to it that it didn't want its victim to see: namely, that it was nothing more than a street-smart liar--a vicious silent killer with a false warmth and a beguiling smile who had only one thing in mind for Cape--his slow and total demise. The little brown bottle had set out to make intimate friends with its intended victim. Now it was only a matter of time. Already Cape's drinking had begun to exhibit an expensive price tag, not just to him, but to those around him.

PASSIONS AND PREJUDICE

You see, Cape had not been the only one in his car the night of the wreck. There was another man in the car with him--Owen Hailey. Owen worked at Spindletop Farms for Cape and me. He was Cape's assistant trainer; and, in later years, he would be inducted into the Horseman's Hall of Fame at the Kentucky State Fair Horse Show. He was just that great of a trainer and horseman. But Owen Hailey was not as lucky as Cape the night of the wreck.

When I heard that Cape and Owen had been hurt that night and that Cape's car had been totaled, I immediately went to the hospital where the two had been taken. Owen had just gotten married, so he was a newlywed looking at a great life ahead of him with a beautiful young bride. When I entered Owen Hailey's room that night, I was not prepared for what I saw. What I beheld in that hospital bed was not the same handsome, dashing horseman that I had known for so long; instead, in his place lay what looked like a twisted, out of place monster from some horror movie. There were only two holes left in his face where his nose had been. It had been literally scraped off, almost completely severed off with the impact of his body being thrown through the windshield into the road and then sliding no telling how far on the gravel and stone. It must have been like having your body sandpapered off with giant sheets of rough sandpaper. His face was a mass of raw tissue. He would carry these scars for the rest of his life."

The doctors said Owen would survive to train horses again. But his face would need endless surgeries and never be the same. That night I told the doctors and the hospital to do everything they could for Owen, and I would take care of it all. And I told Owen not to worry about anything, not one thing. I would take care of his hospital bills, surgeries, or whatever else he needed for as long as it took.

But things are not always as simple and as neat as they could be if others didn't step into the picture. A man came to visit Owen in the hospital.

The man apparently said, "Owen, you're gonna get something out of this for yourself, I hope?" I really don't think that idea ever crossed Owen's mind before. We had been friends for years.

Owen apparently asked the man what he meant.

"Well, that Yount woman has a lot of money. That'd take care of a fella for the rest of his life, ya know."

"It was an accident."

PASSIONS AND PREJUDICE

"Well, sure it was. But that Cape Grant fella was drivin', wasn't he?"

"Yeah."

"Well, he's tied to that Yount woman and all her money--farm manager and all."

"Cape and Mrs. Yount aren't married. I was with Cape in his own car. And we were on our own time. Mrs. Yount had nothin' to do with this. She came up here soon as she heard just to tell me not to worry about anything: she'd take care of it all. She was all distraught and everything. She's a good woman, Mrs. Yount Known her a long time."

"I understand all that, Owen. But you can get more than hospital bills and doctor's fees out of her. You might want to retire some day. A fella's got to look to the future, and you've got that pretty new bride to take care of now. Ya know, Owen, you oughta see a good lawyer and do this thing right."

"Naw, I don't want to do that."

"Owen, things happen. You better get a lawyer after her, I'm tellin' you. At least see one. They'll come right up here to the hospital to see you, 'specially when they find out you're thinkin' of suing that Yount woman. You'll have to fight the lawyers off, if they think they can get a cut of her money."

"I don't know. Look at all she's doin' for me already."

"She doesn't have to know, Owen."

The next thing I knew, I did know. I didn't want to know, but I knew. I wish I had never heard it. But I did. My phone started ringing at Spindletop Hall. It was my lawyer telling me that Owen was seeing a lawyer, and it was all over town that they might sue me. Cape and I weren't married. The car was his. And Cape and Owen had been on their own time. But somehow some little attorney thought he might be able to twist it just enough to get at me and make me his meal ticket for some time to come. It never occurred to me that Owen would try to do such a thing. There was something about betrayal that always mystified me. For thirty pieces of silver, Judas betrayed Jesus and kissed him to boot. How cheap human relations are bought and sold. How easily overturned the strongest of ties can be for a dollar bill. It doesn't take much of a whisper in the ear of a receptive mind to create a mess. Deep down I knew someone had put that blasted idea in Owen's head, or it wouldn't have been there. He was a decent man; he really was. I never knew who the real culprit was. Nor did I understand why Owen had listened to him. But suddenly

PASSIONS AND PREJUDICE

everything was turned upside down. I was hurt and angry that our whole relationship had come down to me being a cash cow for him and some greedy little attorneys who were probably egging him on further. I had tried to follow the Bible's teachings most of my life. I wasn't up to this one, however. As of that moment, I never paid a cent more toward the hospital bills or to Owen. Not a cent. I didn't always do the right thing.

My first reaction was to shut the whole thing off and get away from what had happened. I didn't want to talk about it. I didn't want to think about it. I sported a public lie that I guess came from pure old fashioned pride. I told myself I would give Owen Hailey exactly nothing, not one dime now that he had seen a lawyer. But when I was alone with myself, myself wouldn't allow me to be so dishonest with me. You see, I knew the truth about me. What happened between Owen Hailey and me preyed on my mind; it was a lie to pretend it didn't bother me. I'm sure, truth known, it bothered Owen just as much. But neither one of us would let on to the other. I had wanted to help him at the time, and I would have, but not now. The truth is, what happened between us and how I was behaving as a result of it hurt me so badly I ached inside. I even cried in quiet moments when I was alone with my thoughts. Loosing a close friend is as sad as a death in the family. And loosing a close friend over money is the height of tragedy. *Letting go* hurts, no matter what the circumstances. I was mourning inside, despite what the outside may have looked like. But I somehow couldn't bring myself to do things differently. I kept Owen on at Spindletop Farm. Although on one hand I wouldn't get rid of Owen completely, neither could I forget being betrayed and used by a man I had trusted and considered a friend over the years.

Although we were not married at the time, Cape had a suite of rooms in Spindletop Hall where I lived, too, along with Boo, my help and perhaps my best friend. We originally had this arrangement for security reasons. Spindletop was a big place, and not impenetrable if someone were to set his mind to it. One night, about a year after the anniversary of Cape and Owen's wreck, Cape and I were in the library. The lights were dim. Cape was pouring himself a drink. He asked if I wanted one.

"No," was my unusually quiet response. I watched him lift the glass to his lips and drink down a large amount. Since the wreck, I had been watching Cape repeat this scene time after time. The number of bottles accumulating around Spindletop seemed to be more and more. You could find them in ridiculous places. Boo would be making up his bed in the morning, and there would be a slew of empty ones under the mattress that Cape had been laying on. She would bring them to me. I wouldn't say much.

PASSIONS AND PREJUDICE

Boo would fuss, "Mr. Cape, I don't know how he sleep on all des bottles under his mattress, Miss Pansy. He sure wouldn't be a good candidate for the Princess and the Pea if he cain't feel des bumps under him. Who does he think he's foolin?"

I'd look at Boo, and slowly, almost inaudibly murmur under my breath, more to myself than her: "Himself maybe." And I'd have flashbacks to my first marriage to Albert, a full blown alcoholic. Then I was somehow too young or naive to know what alcohol could do to somebody and how it would keep the prey it was after in the dark, so they wouldn't know what was happening to them. I kept saying to myself, "No, that's not the same with Cape. Cape can go a long time without drinking anything. He can stop." I'd disregard those brief, uncomfortable flashbacks that came as warnings of a hideous time in my life that I never wanted to relive, not even in memory--but I nevertheless did remember, and I remembered vividly.

Cape had a big, beautiful juke box with fancy lights put in his bedroom suite upstairs. Sometimes he would turn it on and just sit there alone in the dim lights of the evening amidst all of his trophies and ribbons, and he'd drink and stare into the empty darkness at something only Cape himself could see. But there were also nights when Cape would drink until Boo and I would get really frightened of him. Boo would come to my room, and we'd lock the big double doors between my suite and Cape's, and she'd stay with me as he pulled on the doors and beat on them trying to get in. Boo and I would listen to the alcohol rant and rave outside, scared he was going to succeed in entering the room. She would hold me, as tears rolled silently down my cheeks, as I once again relived the nightmare that I had gone through in my first marriage with Albert. Boo knew all about that, but I never shared that with Cape. Boo and I told each other everything; she was my confidant, my friend. Then, suddenly, Cape would just shut down, apparently passing out on his bed. But Boo and I would be scared the whole night through. The next morning, Cape would be out working the horses as if nothing, *absolutely nothing*, had happened the night before. Maybe he just didn't remember what had happened. But I did.

On this particular night, as I watched Cape there in the library gulping his drinks, suddenly the words wouldn't stay back any longer. I had kept them in since the time of the wreck. My imprisoned thoughts had become uncomfortable secrets that wouldn't leave me alone. Tonight, they rolled out of my mouth as though someone else were saying them, without softness, without garnishment, as sharp as a razor cutting into a wound to free the pus that kept it sore and alive and secretly hidden below the surface.

PASSIONS AND PREJUDICE

"You're drinking too much."

Cape didn't even look at me. He swallowed slowly. "What are you talking about?" He poured himself another drink.

"Something's bad wrong." He was quiet.

"Nothing's wrong."

"Things have gotten ugly, Cape."

"I don't know what you're talking about."

"Just look at you. Take an honest look at yourself, Cape."

"I'm not hurting anybody, Pansy."

"Cape, last year you and Owen had a near fatal wrecked. You almost died--"

"But I didn't--"

"You disfigured Owen Hailey for life. You were drunk, Cape--drunk and going too fast. That's the reason your car went out of control. You were drinking. You've never acknowledged that--never said a word about the drinking."

"We were having a good time."

BLUE BLOOD

"Is that the way your mind twists it around? Can't you see that same mind might just be lying to you? You were almost killed; your car was totally demolished; and Owen--God--Owen----. You took away a part of that man's life."

"It was an accident."

"An accident that may not have happened if you'd been in full control of yourself. Cape, what's wrong? Can't you see that? In the end it destroyed my relationship with Owen. Every time I look at Owen, I remember. I remember you were drinking and driving and just didn't give a damn about anybody but that bottle that night."

"Well, you won't have to look at Owen anymore." His voice was slow and intent.

I looked at Cape for a long moment, his back still toward me as he drank. "What are you talking about?"

Cape turned around and faced me, leaning back on his elbows on the bar, drink in hand. "I fired him today. A year and a day after the accident. See--," Cape raised his glass as if to toast, "I remember, too, Pansy."

I squinted in disbelief. "What?"

"I got rid of him."

"I told Owen that he could stay on here for as long as he wanted."

"You told him you'd take care of his hospital bills, too, didn't you?"

"That's a private matter between Owen and me."

Drinking with control, he added with a smirk, "A matter of money." He turned slowly toward me.

"It stunk to high heaven when he saw that attorney." Then almost as if talking more to myself out loud, "I offered to take care of everything. Why wasn't that enough?! Why did it have to be *more* than enough? Why couldn't he have just left things alone?" Then I turned to Cape again, "But I never would have gotten rid of Owen. He had a job here for as long as he wanted."

"Well----now he doesn't. I'm the farm manager here." He turned back to his drinking. "Maybe now that Hailey's out of the way, you'll stop bitchin' about that wreck."

"Don't use that kind of language with me. You're drunk. You're the only one who should be gone, Cape." At that point, he turned around and looked at me. I paused for a moment, "Do you really think that you can get rid of your past by just getting it out of your sight? You can't erase your responsibility in that wreck by just getting Owen Hailey out of your way. And you sure can't erase it from my memory."

Cape started to take another drink. I placed my hand over his glass. Then I caught myself, as his eyes met mine. "I'm--I'm sorry," I said removing my hand. "I had no right to do that."

"No, no you didn't." He finished raising the drink to his mouth and gulped it down with a defiant anger only inches from my face.

Regaining my balance, I said with more control, "Cape, you've 'bout near scared Boo and me to death some nights when we had to lock ourselves in a room just to keep away from you when you were drinking. You can't just play like those things aren't happening. And Boo keeps finding bottles everywhere."

"Then maybe she needs to leave Spindletop like Hailey. Cape paused for a moment. "What if I were to tell you, Pansy, that it was either me or her?"

I hesitated. I didn't like Cape's putting me in this position, but I would give him his answer. "Then I'd have to say Boo's been with me a lot longer than you, Cape."

PASSIONS AND PREJUDICE

Cape grew ominously still, his back still toward me. Then with quiet control, while still staring at the liquor in the glass he was turning in his hands, he said as if challenging me in a winner-take-all hand of poker, "You threatening me, Pansy?"

I hated where he was leading us. "You work for me, Cape, the same as Owen Hailey did."

Cape turned and looked at me long and hard, breathing through the alcohol in a whisper, "*The same*, Pansy?" With his eyes locked in mine, he then slowly came over to me and stood in front of me with a threatening closeness, making me uncomfortable as his eyes surveyed my body with an ugly intimacy. "Is it *really* the *same*?"

I turned away from him. "I don't know what's going on with you anymore, Cape." Breaking what seemed to be his momentary hold over me, I stumbled with my words. " I mean, I just want things to be better. You didn't always drink this much." Frustrated, grasping for the right words, I turned to him again, "I mean, is there anything *I* can do to help make things better." Immediately when I said that, I felt myself cringe. My God, what was it that I was saying?! Here I was trying to make things better again, to a lesser degree than I had with Albert in my first marriage--but still trying to make things better by taking Cape's responsibility--a responsibility that only Cape and Cape alone could take.

Going back to his drinking, Cape slurred under his breath, "Take care of your own self, Pansy."

I pulled my words quietly and ever so slowly from the depths of my thoughts and listened to them trail off faintly at the end, "That may be all I can do."

Suddenly slamming his hand against the bar, Cape bellowed back at me, "I don't have a problem, Pansy!"

"Oh, you have a problem alright: that the position of God is already taken and won't be opening up again too soon. Cape--*you're not God*. You can't control everything in this world. You can't accept that. Stomp and rant and rave like a child all you want--worse, like an angry little Rumplestiltskin. But it doesn't change the way things are. I think it's genuinely hard for you to accept the possibility that this stuff could be getting the better of you, and for once you might really need some help with something. ----Cape, there's no shame in that."

"Bullshit. --Oh, excuse me, I forgot, you don't like that kind of language." Pulling himself up, recovering some of his control, "If I had a problem

PASSIONS AND PREJUDICE

with alcohol, Pansy, I'd be the first to know it----now, wouldn't I?"

I looked at him hard and in earnest. "I think you'd be like everyone else with a problem, Cape. I think *you'd be the very last to know.*"

We didn't talk about Cape's drinking anymore after that for a very long time. After all, I told myself, there was a possibility that *I* was the one that was wrong. But that rational line I was feeding myself seemed hollow. I had experienced self-doubt before, and it was almost always the beginning of becoming a victim. People weren't educated much about alcohol then, and I was no different. But if alcohol had been one of Cape's secrets before that night, it had just been given a name and called on the carpet for the havoc it was reaping. It was out in the open now for everyone-----everyone except Cape himself, it seemed.

Wonders never ceased with Cape. The next morning, after that bout in the library, I came downstairs to find Cape sitting in a big, comfortable easy chair reading a book.

I was curious. As I passed him, I said, "Good morning, Cape."

BLUE BLOOD

"Good morning," he said, as if our conversation of the previous evening had never happened.

"Reading?"

"Yep. I've set out to read the Holy Bible from cover to cover. I thought I could do that right fast, but this King James fella is slowin' me down with all his *thee's*, *thou's*, and *doest's*."

I bit my lip trying to suppress a slight smile of suspicion that was creeping across my face. No one could be as disarmingly charming and make a comeback like Cape Grant. I reached into a section of books I had on a shelf, then handed him one. "Here," I said, "This is a more modern version of the Holy Bible. Maybe you can go faster in it.

"Thank you," he said looking up at me with a broad, charming smile. Then, with that same disarming innocence, like a little mischievous boy that was trying to be noticeably good, he started reading again, this time in the easier version.

I left the room thinking: "Do leopards change their spots?" I had always thought they didn't.

PASSIONS AND PREJUDICE

The following year, Cape took the horses to Madison Square Garden in New York. Spindletop Stables set a record that no other Saddlebred stables in Saddlebred history had ever come near, winning three major stakes in one week. Chief of Spindletop won the Junior Stake, the three-gaited championship was taken by Roxie Highland, and Beau Peavine won the five-gaited championship. We also won the roadster stake. So Spindletop had taken the spotlight at the three outstanding shows in the country: the Kentucky State Fair at Louisville, Madison Square Garden, and Chicago. This was quite a feat in Saddlebred circles.

Later, headlines in the *Lexington Herald, Louisville Courier-Journal,* and newspapers throughout the U. S. read:

THERE ARE NO FURTHER HONORS THAT SPINDLETOP HORSES CAN WIN--THEY HAVE WON IT ALL

And indeed we had won it all.

The riding lessons I had started with Cape in Beaumont had continued through the years under the direction and perfectionistic eye for detail of the master himself, Cape Grant. My skills were

BLUE BLOOD

forever being honed with each new day. They say that the student will one day surpass his teacher. Not in this case. Cape was the undisputed king of the show horse ring, and, would forever be. I would never be that good--but that was not to say that I was bad--certainly I had become over the years no slob at riding and showing. There was something about my sitting atop a Saddlebred and going through their moves with them with focus and total concentration that was like partaking in holy communion. It gave me a thrilling sense of total freedom like none other, like being in church and feeling the hand of God move over you in a miraculous way--there was an ecstasy about it at certain moments. It was a feeling of empowerment like you were a part of nobility, sharing royal moments, the regalness of which supposedly only belonged to kings and queens. The sense of accomplishment you experience is like no other. There was always something to improve on, always something to try to do better-- a constant challenge--a challenge that is only for those brave souls who like to climb mountains in life and feel the victory and exhilaration of reaching the peak where there's no one but you and God and the beauty and force of pure, determined skill. To ride atop a Saddlebred is an honor and gift that must be a little akin to being knighted. At least that's the way I felt about it. *When I rode an American Saddlebred, I felt genuinely beautiful, inside and out.* I loved those animals.

PASSIONS AND PREJUDICE

I was more than fully capable of dismounting a Saddlebred horse by myself after all these years, but often Cape would rush over and put his hands on my waist like a guide that was pointing me back down to earth after a ride. His hands would linger on my waist as I touched the ground--for what seemed all too many moments. And when I'd turn around, his eyes would be looking at me with an intensity and in a way that was uncomfortable. And there was that awkward pausing--always that awkward pausing---after which he'd ever so slowly remove his hands from around my waist. I'd hand him the reins, and leave the barn in the self-consciousness of that moment, in a more hurried manner than normally, it seemed, trying to act all the while as though I saw nothing that was happening between us. My face would feel flushed and hot. And as I walked back up to the house, I could feel his eyes watching me--following me and my every move--for all too long.

Spindletop Hall was at last completed in 1937. I had inscribed in the mantel of the oak room, **"East, West, Home's Best."** I loved that place. No place in Kentucky could compare or was even in the running with Spindletop Hall, no place in the entire U. S. for that matter.

BLUE BLOOD

I had all the furniture in Spindletop made to scale. And I had the gardens both in front of and behind Spindletop modeled after those at the Palace of Versailles in Paris. They were breathtaking and immaculately taken care of at all times. There were big spreads done in the papers about Spindletop Hall during this time:

Lexington Herald-Leader, September 18, 1938

WITH WORLD TO CHOOSE IN, MRS. YOUNT PICKED LEXINGTON

SPINDLETOP HALL PALATIAL HOME

Planned for Comfort, House is Beautiful as Fairy Princess'

Not mentioned in the newspaper articles was my blue and gold macaw parrot that used to scream at people who came into the house like a self-appointed watch dog. My parrot would not have liked to have been left out of the papers. I was careful not to tell him, too, since macaws have the understanding of a three-year old child, and he would have probably taken exception. Another thing not mentioned was that I liked to cook as much as Will and the other chefs I had on staff.

PASSIONS AND PREJUDICE

Cooking was my hobby. And Will, Boo, and I were always exchanging recipes. One of my favorites was "Sunday Night Chicken Sandwiches." For that you steam one chicken covered with bacon and butter. Then cut it up fine and add sweet red peppers, sweet green peppers, and mayonnaise, and spread between your favorite bread. And to "Pansy's Famous Homemade Chili," I liked to add lots of sweet potatoes. Nothing's more healthy than sweet potatoes. For turkey or chicken dressing, I liked to "sweeten the bird" by adding raisins to the dressing. If I were making a cocktail sauce, I would often mix catsup, chili sauce, celery (real fine), chives, and bell peppers to taste. Boo had a good recipe for stuffing vegetables like squash: she'd make a cream sauce and chop up shrimp and onions real fine in it. I gave special attention to setting up the kitchen at Spindletop Hall.

I had literally poured bottomless wealth, and more importantly, all my heart and soul into the creation of Spindletop Hall and Spindletop Farms. It was my passion. I wanted it to be a showplace for Kentucky that everyone could be proud of. And I would have welcomed anyone with open arms to share in my joy and in my dream.

Spindletop Hall was finished and our horses had won everything there was to win, just as the newspapers had said. And now I turned my

BLUE BLOOD

attention to breeding, with the same determination that I did everything else.

My first dream was to breed Roxie Highland and Beau Peavine, two of the all-time greats in Saddlebred history. I had a real soft spot in my heart for Roxie Highland, both because I had bought this champion originally for Mildred; and, secondly, she was not only beautiful, but she was one of the greatest show horse winners of all times. Roxie had a type of attitude about her. She had a soul of pure fiery temperament, and there was something about the way she "wore her ears" and the "direct eye contact," so to speak, that she gave the audience that made her have an unparalleled, fascinating presence. Her debut at Madison Square Garden was a classic, as reported in the *New York Times*:

WINNER IN 1934 WITH $253,500

ROXIE TRIUMPHANT IN DEBUT

Temperamental Three-Gaited Mare Is Queen of Ring at Garden Show

New York, take a long, comprehending look at Roxie, one of the greatest actresses that ever infuriated

PASSIONS AND PREJUDICE

the horse show judges! She is Roxie Highland, queen of Mrs. M. F. Yount's Spindletop Stables' show string, here from Beaumont, Texas. Undefeated in four years of campaigning in the heart of saddle horse country, Kentucky and Missouri, this great three-gaited Missouri-bred mare came to New York last night and conquered in her debut at the National Horse Show.

Exceeding her advance notices, Roxie exuded temperament. She ignored the other horses in the ring. She calmly stared at dowagers and magnates as she passed. Then, swinging out of the line for the climactic "show" in her debut class, she trotted with a kinship to Pavlowa and Karsavina, queens in their own right who captivated with rhythm, grace, and fire....

At Louisville, Chicago, and St. Louis the Spindletop horses have been cutting a wide swath through the saddle classes. Roxie Highland again won the great walk-trot championship stake at Louisville, so the cavalcade moved on New York....

BLUE BLOOD

Roxie won at Madison Square Garden in New York and was retired there in 1935 in one of the most beautiful retirement ceremonies ever attended. Cape Grant rode her at the ceremony, showing her at her absolute exhibition best. It was hair-raising as she strutted her stuff. Newspapers worldwide called that exhibition "the icing on the cake for an outstanding career in the ring." Roxie's retirement was carried nationwide on radio. Cape and I stood beside her in the ring and accepted a floral horseshoe and some silver. Representatives from the city of Beaumont, Texas were there and presented Mildred with a silver trophy honoring Roxie for her distinguished career and for being such a good ambassador for Beaumont through the years. The grandeur of the ceremony in its tribute and music to Roxie over the loudspeakers, and the special retirement blanket of white lilies put over her that night as she walked before that packed, standing ovation crowd of her applauding, admiring followers, gave the feeling of the entire Saddlebred world paying homage to true royalty--a queen in the realm of noble blue bloods whose skill in the ring, unequivocal beauty, and lifetime of contributions and accomplishments were being celebrated that night to the full extent of human ability and to the full extent of the human spirit. As I listened to the announcer's voice reading off Roxie's career accomplishments and the accompanying full swells of *My Old Kentucky Home* being played by the band, I found myself caught up in a flood of tears and emotions, and applauding as wildly in respect and appreciation as anyone there. I was proud for Roxie, and proud to be a part of the Saddlebred world. She was proud, too, I think.

But our breeding Roxie to Beau Peavine was heartbreaking. The first foal that everyone was looking forward to with great anticipation died of pneumonia. That was 1937.

1937 was also the year of the inception of the Junior League of Lexington. I was one of the original supporters and "pushers" for the organization, and I gave generously to them since I felt it would be a key organization in the further development of American Saddlebred horses, as well as good for Lexington and Kentucky.

During these years, Mildred would visit me in Lexington during the summers and on holidays from school. The bodyguards were as much on duty now as they were when she was thirteen, which still made dating for Mildred a little on the abnormal side. But it was necessary for security reasons. I remember, in particular, a handsome young man that I had heard of in Lexington who took Mildred out on a date at my request. His name was George Stanhope Wiedemann, known as "Hope" to his friends. He came from a good family, and he had quite a reputation with the girls. I think Mildred genuinely enjoyed that date with Hope because it was different, and rather imaginative.

BLUE BLOOD

Stanhope gave a report of the date to a friend which I later read. It sounded right down Mildred's alley to me:

> The phone rang just before I was going back to school at Lawrenceville in September of 1936. The caller introduced herself as Mrs. Miles Frank Yount's social secretary. She wanted to know if I had a date New Year's Eve. At my age, and with it being only September, of course I didn't. So I said so. The secretary then asked would I accept an invitation to escort Miss Mildred Yount to the New Year's Eve Ball at the Lexington Country Club. I stammered that I would be delighted, being somewhat overpowered by the aggressive secretary. She then invited me to meet Mildred on Saturday morning and gave me the time of 10:00 a.m. The place was familiar.
> The meeting place was a farm house on the southwest corner at the intersection of the Ironworks and Newtown Pikes. When I worked on the farm which was then Shoshone Stud, owned by Mr. W. R. Coe of New York, I boarded there. Mr. Hugh Fontaine, then Farm Manager, gave me the summer job of joining eleven colored boys to break yearlings. When I became too heavy to ride after one summer, he assigned me to the "muck truck." I became familiar with the farm and the employment of

153

the barns. My room in that house was the upstairs left bedroom. Years later when I met Mildred, the Younts, Mildred and her mother, were living in the house while the huge mansion Spindletop Hall was being built. Needless to say, the house had been "fixed up."

Mildred and I met in the living room which had been enlarged by knocking out a partition between two rooms. After being admitted to the house at the appointed time, I was ushered into the living room to await Mildred. At the end of the room there were three chefs all wearing white toques, behind a table set with caviar and champagne. My doubts about too much formality were dispelled when the secretary introduced me to Mildred who was a very natural, sweet person. We had no strain during this short encounter. I looked forward to the New Year's Eve Ball.

When I came home for Christmas vacation I called to confirm the big date. I was given the option of having a chauffeured limousine for the night, or I could use my car. I chose to use my car which was a well used Chevrolet sedan shared with my brother and sisters. I remember putting a clean blanket over the front seat to protect Mildred's dress. Mother said, "If you go clean and mind your manners, it makes little difference what you drive."

BLUE BLOOD

The Younts had moved from the farmhouse into the mansion when I called for Mildred New Year's Eve. We met in the front hall and she was ready to go. She had on a corsage of white orchids which made me swallow because I had brought her a modest corsage of white camellias. With no hesitation she dropped her orchids into the umbrella stand and put on my lesser corsage, despite my protest. We departed for the Club.

I noticed when I pulled out of the driveway that a large black Buick, manned by two men, followed discreetly. I asked Mildred about it and she explained that "Mother" was concerned about her safety and insisted on the body guards, much to Mildred's embarrassment. I told her we might have some fun with them.

The Ball was enjoyable for her because I was able to introduce her to quite a few boys and girls in the crowd. The boys did their duty dances which kept Mildred busy. She liked that.

I hadn't seen the body guards, but at midnight when the lights were momentarily turned off to herald the New Year, I felt a slight nudge and the lights were turned back on. The body guards had sandwiched Mildred between them. When they saw she was alright, they melted into the background. When we left the Club, I asked Mildred if we could try to ditch them. She was skeptical

PASSIONS AND PREJUDICE

because "Mother might not like it," but she said to go ahead.

Instead of turning left from the Club gate to go into town, I turned right on the Paris Pike and immediately took the first turn off the Paris Pike to the left which was Swigert Lane. I pulled into the first sheltered driveway on the right and shut off the engine and lights. The big Buick roared past, going down Swigert Lane. I backed out and returned to the Paris Pike where I encountered a stream of cars coming out of the Club gate on their way into town. I was held up until Garland Barr recognized me and paused long enough to let me enter the flow into town. Apparently, the body guards had no such luck. Mildred and I lost them completely.

Rather than go to the Question Mark Cafe on Main Street where everybody met after a dance, I took Mildred to our house where our cook Mamie had just baked an angel food cake. We had the best milk one could get, it being University of Kentucky Experiment Station milk which was raw, but it was Bang tested daily. We enjoyed cake and milk and a private conversation.

I took Mildred home at a reasonable hour, and we said our goodnights. Subsequently, I was invited to supper at Spindletop which was delightful. I made a hit with Mrs. Yount when I complimented her

BLUE BLOOD

Remoulade sauce which she made from shrimp.

Indeed, Hope Wiedemann did make a hit with me. He was a most remarkable young man.

Mildred often went to our summer house in Rockledge, near Manitou, Colorado, in the summers, as well. Rockledge was beyond all description, located in the mountainous Pike's Peak area. Frank and I and Mildred used to spend many happy days there together in the summers, and sometimes at Christmas. It was a fairy land. Rockledge seemed to grow right out of its naturally beautiful setting of mountains and native forests of pines and cedars. There was a view of Pike's Peak out the front of the house through the mountains; and to the rear of the house there was a breathtaking view of the Garden of the Gods. To the east were natural terraces leading to a valley below that could have only been decorated by God himself with every possible type of Colorado wild flower. There was virtually nothing left of beauty that one could possibly desire in a landscape, so incomparable was this area in natural beauty.

The people of Manitou and Colorado Springs were also incomparable in their attitudes toward people from other places, I thought. I remember one lady in particular--Mimi. She went out of her

PASSIONS AND PREJUDICE

way to show me around, to direct me to the best places to buy this and that for Rockledge, even taking me around in her run-down, old car that coughed and sputtered all the way; she wasn't even willing to take anything for the gas it cost her. Mimi even offered to share her bagged lunch with me, a perfect stranger; I was touched by her kindness. She had such an open face and just seemed to be in love with life itself. We laughed most of the time we were together. And although she obviously didn't have much in the way of money and material things, I can tell you she was richer than many that have it stuffed under their pillow cases. I couldn't help but remark to myself that wouldn't it be a better world in every way if we all went out of our way for each other and placed the welfare of others above our own, like Mimi had done for me; and wouldn't it be a better world if we all treated each other as royally as Mimi had treated me, a stranger to her, no matter what we had in money and possessions. This woman, without knowing me from Adam, had treated me with such loving care that when I returned to Lexington, I went out to a Ford dealership and bought a brand spanking new red Ford car with all the trimmings and had one of our farm managers, Ed Fitzpatrick, drive it out to Mimi in Colorado Springs, and give her the keys, the title, and this note:

BLUE BLOOD

*Thank you for simple human kindness,
The stranger you helped from Lexington.*

[signature]

Except for that one brief encounter, I never saw Mimi again. I was told she died a year or so after that; but many a stranger shared Mimi's hospitality in that shiny red Ford car before her death--and Mimi laughed the whole time she was driving, with her usual joyful outlook on life. She never changed a bit from her down-to-earth manner--the only difference toward the end was the car. It didn't sputter and cough anymore; and I was told she would rather have lived in that car than at home. Thanks to Mimi, both Colorado and Rockledge and its people took on a special aura to me from that visit forth.

I know what some are thinking: but what on earth possessed her to give that woman a car? Everything is proportionate; and I never believed that kindness should be taken for granted. I believed that if you were in a position to do so, kindnesses should be rewarded many times over with the hopes that the doer of that kindness would be encouraged to continue on and touch others in this life. I have no doubt that this type of thinking seemed a little "crazed" to some people, especially since money often rivals God

PASSIONS AND PREJUDICE

Himself in importance, if not surpasses Him. But it made perfect sense to me. I gave because it was the "right" thing to do, by my way of thinking. I had noticed in my life, long before I had anything, that some rich people got a type of perverted pleasure or feeling of power out of getting someone in obvious need to ask them for money and then turning them down flat. Frank and I had gone through that many times in trying to put together the money for drilling on Spindletop Hill. I loathed such behavior--people who purposely humiliated others who were obviously in need of either an opportunity or money. I never got a kick out of watching anyone suffer. Such people seemed to forget the all equaling power of life's course-- namely, it could change on a dime. The one that was up today could very well be the one down tomorrow, and vice versa. I promised myself that if ever I had anything, I would make a conscious effort to do some things differently than some of those that Frank and I had run across when we first started out.

Too, I guess it went back to a homily Father once gave at St. Anthony's Catholic Church in Beaumont when I was a kid. I remember it was based on Luke 21:1-4. Father looked down at us with his intense blue eyes that Sunday and said that in Jesus' time a lot of rich people were putting their gold and silver into the offering plate and patting themselves on the shoulders before one another for their generosity. Father said what these people were trying to pass off as generosity was a fraud, nothing more than greed cleverly disguised to cover the true self-serving nature of their gestures--gestures contrived to feed an ugly hunger for attention and praise in their empty souls, souls that had no idea what love really was. Father said we should feel for such people, that they were indeed sad. But Jesus, he went on, noticed a poor widow who only put two small pennies in the offering plate. Jesus said the rich ones who had more than enough really gave little in proportion to what they had; but He said this poor widow, who had less than enough, actually gave the most because those two pennies were all she had to live on. Everything is proportionate to what we have, he said. So, a car was nothing for me. I had a long way to go if I ever planned to equal that widow's true generosity and giving.

Maybe it was remembering Father saying to us as children, "It's all proportionate. We have a duty to others less fortunate, depending on God's portion to each of us." Maybe it was my Catholic upbringing, in general, or maybe it was because I knew first hand how it felt to walk in the other person's shoes; or perhaps both. I don't know. I simply tried to do what I could, where and when I could, remembering God's infinite mercies to me, one who had once had less than nothing. Mimi caused me to reflect a lot on life, where I came from--which never really left me in my life, I think-- and the way I thought about money, and

PASSIONS AND PREJUDICE

possessions--which were never that important to me--and happiness--which was everything, along with health. In that respect, you could say that sweet little unknown Mimi, who was mostly lost in the crowd of the human race, as far as fame and status and worldly goods were concerned, made a rich difference in at least one life--namely mine. I hope I made at least some difference at the end in hers.

There could not have been a more perfect place to fall in love than Rockledge. It was like falling in love in the midst of the grandeur that must have been God's original Garden of Eden. Rockledge was a type of Garden of Eden for special love, with a purity of beauty that defied all description.--Words failing--lets just say, unlike Kentucky, it was a place without thistles.

It was in Manitou on one of her summer visits to Rockledge that Mildred met Edward Daniel Manion, a young Tulsa, Oklahoma attorney, then vice president and general manager of The Sinclair Refining Company, and pipeline department. There they fell madly in love. And they married. We held the ceremony in Lexington at St. Paul's Catholic Church where I always sat in the third row from the front on the right-hand side every Sunday for Mass. We had the reception at Spindletop Hall. It was a wedding and reception, the grandeur and beauty of which would have rivaled and been the envy of royalty anywhere; in particular, I supp[ose] it was the envy of that royal old society that wa[s] Kentucky during these times. Newspaper spreads appeared in the *Lexington-Herald* and the *Beaumont Enterprise*, with titles like the one that appeared in the *Tulsa Tribune* on Monday, June 27, 1938:

Mildred Yount, Southwest's Richest Oil Heiress, Marries Young Tulsa Attorney in Lexington, Kentucky

Suddenly, Mildred was gone. The one aspect, now reality, of the wedding that my mind did not want to acknowledge was happening. I felt the bottomless emptiness that was probably always there waiting to come out, that was underscored all the more by the whirlwind of activity that surrounded Mildred's wedding. I was alone, and I felt the deep, painful caverns of loneliness that make-believe joyous activity and lots of people around deny so well.

I had pushed to go to the top of the Saddlebred industry in a foreign country called Kentucky, that was Frank's and my dream together. I did it, but without Frank by my side. I now had to face one of life's scariest specters that had been waiting in the shadows of Spindletop for me-- namely, letting go, a ghostly premonition of what is

PASSIONS AND PREJUDICE

yet to be in each of our lives. Then there came that famous drivenness and flurry of activity in building Spindletop Hall that helped me to forget Frank's absence for a moment, like drinking a hard liquor, and pretend that I was building for all of us once more as a family, and for Mildred most of all. Mildred and I had moved into Spindletop Hall, which was a dream realized, in 1937; and Mildred was excited about her new room and suite that we together had spent so much time planning and furnishing upstairs. Somehow, that too, like Frank, I thought would last forever. But things change. In less than a year, Mildred was married and gone. But I was still there, alone in a monstrosity of a mansion so large one person could get lost in it. It seemed I was inevitably out of step with life and always behind life's ever-onward-marching-and-never-looking-back itinerary. I was learning again about letting go. I had pushed and pushed and pushed. And I had built and built and achieved and achieved and accumulated and accumulated. The irony was that I was no less alone in Spindletop Hall and on Spindletop Farms in Lexington than I had been as a child sitting alone by my mother's unmarked grave in Magnolia Cemetery, crying in my coarse apron as the other kids poked fun at me for the way I dressed and the glaring fact that I had nothing. Now, I supposedly, in life's terms, had something--according to some, I had everything there was to have--but I found deep down that there was no difference between everything and nothing. I was still me alone with me.

BLUE BLOOD

In 1938 I held the first annual Spindletop yearling sale, and it revolutionized the Saddlebred industry. For the first time, the general public had the opportunity to buy the foals of some of the greatest show horses in history--foals sired by such greats as Beau Peavine, a stallion beyond price who had already proven he could pass on his exceptional qualities to his get. The Spindletop Yearling Sales were a needed innovation in the Saddlebred horse industry, and they set the pace for other saddle horse nurseries to follow. In other words, I challenged the industry, which I thought needed to be done for the sake of all horse people. In 1938, *Saddle and Bridle Magazine,* "the oldest name in show horse magazines" in the U. S., carried this article:

Spindletop--Ace of Saddle Horse Nurseries

Visiting Spindletop and the foal of champions brings an astonishing realization to mind: in producing youngsters of this type, Spindletop is doing an enormous favor to the show horse public, for in giving them the opportunity to buy this type saddle horse either privately or at auction in their yearling sales, they are doing

PASSIONS AND PREJUDICE

> *immeasurable good for the show horse game.*
>
> *When Mrs. Yount and Manager Cape Grant decided that Spindletop would be a breeding establishment, they made up their minds that it would be one of the best in the country--and so it is.*

But again--so what? I was climbing mountains, reaching the peaks, and finding out when I got to the top that what I wanted most was not there. Instead, there was only thin air, which couldn't hold me, or comfort me, or warm me on cold, lonely nights. Ever try to get in bed with a bunch of trophies, and ribbons, and newspaper clippings singing your praises? None of them makes good foot warmers.

Alone, indeed. I had a small handful of servants who were my friends, and, I liked to say, tongue in cheek, I got lost at least once a day in the opulent catacombs that were Spindletop Hall.

I had not received invitations to darken the doors of the Lexington Thoroughbred people or its privileged old society, but I thought that shouldn't keep me from inviting them to Spindletop Hall. So I decided to give a party. If Kentucky would not come to me, I would go to Kentucky. I wanted them to know the doors were open, and I wanted to be their friend. I would welcome them with open arms as was the down-

BLUE BLOOD

home way in Texas. There you never met a stranger. And I had never met one from my side, although it seemed that Kentucky blue blood ran a little cold, and they were perfectly happy with my remaining a distant stranger in their books. I was determined not to oblige.

I sent out invitations to all the Thoroughbred and society people to come and enjoy my Spindletop. I imported wines and liquors from all over the world, and had the best of foods brought in and prepared in the finest ways by our chefs. I had even had that dance floor at the bottom of Spindletop Hall built special with a "give" in it so when we did have a party, when people danced, their feet wouldn't get tired; and they could dance and dance and dance all night long. I hired the best country western band I could find for the evening. Since I had gotten a couple of Arctic blasts from some of these folks before, I figured some real Texas "Yee-ha" music would help break the ice and loosen everyone up.

I was upstairs dressing when Boo, my maid and friend, who had been with me from the beginning in Beaumont, came in. She was much more like family than just a friend.

"Miss Pansy, all the food's been set out and arranged, and we *is* ready fo' 'em."

"Thank you, Boo."

167

PASSIONS AND PREJUDICE

"Miss Pansy. You ain't gonna wear that dress to yo' nice party, is you?"

"Why," I said, looking down at it and surveying it once more. "What's wrong with it?"

Going straightway to the clothes closet, "Why, it's too plain, Miss Pansy. What about this shiny red dress you bought and never wore."

"I'm not anything but myself, Boo."

"But for a party like this with these highfalutin society folk, you needs to sparkle a little." Boo took the sequined red dress out of my closet, a little like an old mother hen.

"I would just look silly, Boo. Pretentious."

"Miss Pansy, these pretty dresses just go to waste in that dark closet. And what about that pretty necklace you have in that vault. Time to bring it out and shine, girl."

"Boo."

"No'm. I can't keep my mouth shut no more."

"Did you ever, Boo?"

"Miss Pansy, now you listens to Boo. Ever since Mr. Yount died, yous never tried to dress up no mo'. You and hes used to close those places down, dancin' in Texas. An' you sparkled. This is yo' comin' out party in Kentucky." She handed me the dress. I looked at it, then her.

"You're gonna be the end of me yet, Boo."

"No'm, I'z gonna keep my promise. I'z goin' to fix yous up. It's about time somebody told you. Now put that dress on, honey."

I made a face. Boo was a hard one to combat once she had her sights set on something. And I knew she always had my best interest at heart. "This once. But we're not going to let this become a habit, understand. I'm not comfortable in this kind of showy stuff, and I'm not like them people anyway. They were *all born* with silver spoons in their mouths."

I unzipped my rather plain, I admit, print dress and slipped it off. I looked at Boo who was radiant with the biggest smile you ever saw.

"Well, put it on, girl." I slipped on the red sequined, tight dress, and Boo zipped the back for me. "Oooo--weee! Now that's a party dress."

I looked in the mirror at myself. "I look like a cow."

PASSIONS AND PREJUDICE

"No'm. Dat's not so. Now where's dat sapphire and diamond necklace yous has? That's pretty 'nuf fo' da' Queen of England herself."

"Then the Queen should wear it. It's gaudy."

"Now, Miss Pansy. You stops being contrary 'bout dis and take dat necklace out of dat vault. It belongs 'round yo' neck, not in some dark hole."

I reluctantly took out the necklace under Boo's watchful eyes, and put it around my neck. Boo snapped it to, "ooing" and "ahhing" all the way.

Boo slapped her leg, chuckling. "Miss Pansy, just look at yo'self. Now you look like you's havin' a party."

"Like I said, Boo. Only this one time, understand."

"I'm gonna go downstairs to make sure Will is ready. Now you come on down and greet all those guests. Theys should be arrivin' real soon."

"I'm gonna stand at the door, Boo, and tell every one of them, I'm glad they come.--I hope they like the music and food. I want Kentucky to have a good time at Spindletop tonight."

"Ain't no way they won't with yo' hospitality and all these fixin's, Miss Pansy. Now don't you

170

worry yo'self. It's time yous had a good time, too, for a change."

When I came downstairs, everything was in the best of order. The servants were ready, the doormen in place, and valets outside; and Spindletop was lit up like a Christmas tree to welcome its guests with joyous arms. So I waited. And I waited----and I waited--------but, no one came. The clock sounded the hours that past. I walked to the open door more than a few times. Yes, we waited. But, indeed, no one came. No, not one. The servants and doormen cut their eyes at one another awkwardly, not knowing what exactly to do under these strange and strained circumstances. I poured more than one drink of the finest Kentucky bourbon money could buy. And we dared to look at one another, at which point, out of an embarrassment that no one wanted to share, we looked away almost equally as fast.

Boo was always the brave one. After what seemed like an eternity of waiting and watching, she came over to me as I gulped down another drink, and she put her hand tenderly on my arm. "I'm sorry, Miss Pansy." We looked at each other directly in the eye for the longest, as if searching for some explanation that would make things

PASSIONS AND PREJUDICE

alright. "Miss Pansy, I don't think they's comin', honey." Her low, kind voice jerked me from my denial. She was right, and I knew it. They weren't coming. No one. Not one. It was an unprecedented slap in the face, and it just as well to have been administered to me by a very big sumo wrestler.

As is often typical when the unthinkable happens, I found myself suddenly trying to deny I had just had the wind knocked completely out of me by some invisible little social fist that was probably wearing a proper white glove. I found myself trying to make the best of a god-awful situation with an obviously false cheerfulness that was so transparent it only added to the embarrassment: "Well, so much food. A wonderful band. What are you all waiting for. Can't go to waste. Everybody dig in."

The servants and doormen looked stunned at my request. Then Boo took over as if trying to smooth things over that would never be the same again, "You's heard, Miss Pansy. Everybody grabs you a plate and eats ya'll some of this fine grub. It done come from all over the world. Like Miss Pansy says, wes don't wont to waste it. And ya'll over there. Mr. Band leader, you and dat band play somethin' for Miss Pansy, seein' you's bein' paid for the ev'nin' more than you'd normally be gittin' in a month of Sundays."

BLUE BLOOD

I barely heard any of this, I was so numb. I think I mumbled, "If ya'll excuse me, I think I'll go up to bed now." I remember almost clinging to the railing for extra security on the winding staircase as I went slowly up to my bedroom. All the servants seemed caught awkwardly between Boo's rushing around trying to get them to fill their plates as if nothing was happening, and their daring to peer on the sly after their fully humiliated lady of the house taking her leave. At that moment, it didn't really matter to me what they were looking at or thinking.

I closed my bedroom door behind me, and leaned with my back against the door inside. I undid the necklace fit for a queen and let it drop to the floor; and I unzipped my "shiny" dress that Boo had thought was so wonderful and let it fall beneath my feet. I stood in my slip, peering into the emptiness. I mechanically poured myself another strong drink. As I drank, I was suddenly startled by the band downstairs striking up--of all songs--*San Antonio Rose*--so loudly that they just as well have been playing in the shadows of the corner of that softly lit bedroom. *San Antonio Rose.* The liquor intensified the flood of memories that were crowding in. *San Antonio Rose* had been one of Frank and my favorite songs in Texas. I remember how we had danced that night at the Texas State Fair under that big Texas moon before Spindletop changed our lives--just him and me, as

PASSIONS AND PREJUDICE

if we were the only two people on that dance floor.

    My eyes fell on Frank's picture on my lamp table. Frank in that white linen suit that always made him more striking than striking. And as that haunting melody swelled, I turned around. And it was like I suddenly saw Frank in the shadows of that room with me. I stopped, almost holding my breath, then I dared to call his name faintly in a whisper, as if afraid he might go back into the shadows and leave me. He said nothing. But he seemed to step out of the shadows toward me. He took me in his arms once again, looking deeply into my eyes, and then pressed his soft, warm cheek against mine. I closed my eyes. We danced. We danced once more as we had so long ago under that awe-inspiring, big Texas moon, as the strains of *San Antonio Rose* echoed through the corridors of the dream that we had once shared together, melting our souls passionately into one out of kindness and mercy, just one more time. Then the music stopped abruptly, as startlingly as it had started. There was a deafening silence. And Frank was gone. I called his name, picking up my drink and gulping it this time like an anesthetic. As I turned, my tear-brimming eyes caught my reflection in the full-length dressing mirror off to the side. I didn't like the blurred figure I saw reflected there that night. It was as if I had lost my definition in life. Nothing seemed clear anymore. I threw my glass, shattering both the glass and the mirror. And I cried out "Frank, Frank!," collapsing across the bed. And I sobbed, calling his name in the pain of the loneliness and abandonment I felt to the very depth of my being.

The next morning, when I came down, everything had been neatly put away, as if nothing had happened. I crossed the room going toward the kitchen. That's when I overheard one of my servants talking. It seems they had heard the maids of some of Kentucky's most illustrious society gossiping on their way to work that morning. What Kentucky society was saying was this: "Her money can buy her a lot of things; but the one thing it can't buy her--is "*culture.*"

That sort of said it all.

But if that wasn't the kicker, some of the Thoroughbred people who wanted to be even cuter were sniggering behind their dainty white gloves over tea and crumpets: "Too bad she didn't have the breeding to put R. S. V. P. on the invitations. Then she would have known we all weren't coming."

Despite all the things that happened, however, it didn't keep one of them from coming

PASSIONS AND PREJUDICE

to my door for donations. I still remember the two ladies from the Junior League of Lexington who came by. As I said before, I believed in the Junior League; I promoted it heavily because I believed it was good for Lexington and Kentucky and the Saddlebred horse business--and it was; and I gave to it generously and gladly. I was, after all, one of the first to push for its existence.

Spindletop Hall was so big that I had had intercoms put in every room of the house so the servants and I could find each other without having to whistle at the top of our lungs, which I did sometimes resort to since I grew up with some mighty powerful whistling champs. I remember once, before we got the intercoms, a decorator from Lexington was out helping me fix up Spindletop Hall. I whistled from the front foyer to the back kitchen for a notepad, thinking nothing about it. When I turned around and saw the decorator's face beet red, I thought the man was going to expire. I was just being practical: why walk all that way if you can whistle. Anyhow, the point was that I could hear what was going on now in any room of the house after we installed those intercoms and had them turned on.

The two ladies from the Junior League of Lexington standing before me that day said they were hoping I could contribute again to their fund drive. I was happy to do so. I told them if they'd wait in the foyer, I'd go upstairs and write them a check and bring it down. I was thinking of a six figure number for the League.

As I sat at my desk upstairs and was writing them what I thought was a generous contribution that would help, I suddenly heard two voices coming in loud and clear over my desk intercom that one of the servants must have left on by mistake. The two ladies were seemingly, in my absence, helping themselves to a grand tour of Spindletop Hall. They seemed to have come to a halt in my new bath where I had just had the latest French bathtub imported and installed. The two ladies were commenting uninhibitedly, when suddenly one of them said to the other, "Do you really think that old walrus would fit in that tub?!"

Old walrus! Old walrus, I thought. I tore up the check for six figures, and started writing a new one, this time with one figure. And I signed it, "Pansy Yount, The Old Walrus." Then Lucky, my little Pomeranian who thought he was a heavy-of-a-guard-dog, and I carried the check out to the stairs. The ladies were back in the foyer, good as gold. They looked up and waved and smiled to me. I waved down, and smiled to them. Then Lucky and I went down the winding staircase to where they were standing by the front door. I handed them the check. They looked horrified as they looked at it. "But, Mrs. Yount," they said, "can't you give more to the drive?!"

PASSIONS AND PREJUDICE

"Ladies--the 'Old Walrus' doesn't feel like giving more today." They practically stumbled out the door as Lucky, sensing it was a moment to seize, suddenly turned from a Pomeranian into a Doberman, baring his teeth convincingly. That ended my association with the Junior League of Lexington. That incident went though Lexington like a shot! As they say, one bad apple can ruin the barrel. Well, here were two. But I still believed in the purpose of the League itself, I just didn't donate anymore for a while out of respect for myself. I knew, too, that some day--maybe not in my lifetime--its face would inevitably change--and for the better.

Something else had been happening during these years between 1934 and now, since I turned my attention to Kentucky and securing property there, building Spindletop Hall outside of Lexington, and Spindletop horses cutting a wide swath through the Saddlebred industry. It was another one of those *13ths* that so haunted my life and that of Frank's while he was living--another of those mysterious *13ths* of Spindletop that had crept up on me while my head was turned the other way and I was busy with Spindletop Hall, Kentucky, and the horses. I was so busy that I had not noticed something that had caught on in the newspapers. Specifically, in the "funny paper" section. It was a comic strip that was given birth August *13th,* 1934, at the very same time I had started looking for land in Kentucky and had

BLUE BLOOD

begun planning Spindletop Hall outside of Lexington. By now, 1938, the comic strip was gaining momentum all over the U.S. and was gathering an international audience, as well.

I remember how it came to my attention. I came into the kitchen at Spindletop Hall one morning and saw some of the servants laughing. When they saw me, they quickly dropped their mirth, and put something behind their backs. Now these servants had been with me a long time. They were, for all practical purposes, my Kentucky family and friends, and I treated them as such, much to the consternation of many of Kentucky's old society. I gathered they thought I was setting a bad precedent with my servants before their servants. Kentucky's old society, with some exceptions, seemed to operate more on the basis of an old feudal system where servants were, for all practical purposes, slave-like creatures that they thought they could talk to worse than their dogs. But I didn't care what they thought. At Spindletop Hall, the servants and I often ate together in the little kitchen, chatted, and shared stories. So I knew their expressions well. And I also knew when something was up. And something was up.

"What's so funny," I inquired. They all suddenly looked like they had canary feathers hanging out their mouths, a little like my cat when he was struck with guilt over being caught in the act of doing the unthinkable, except to a cat.

179

PASSIONS AND PREJUDICE

"What's that behind your back, Sissy," I asked one of the cooks.

"Nothin', Miss Pansy. Nothin' at all," came the reply. And everyone was suddenly busier than busy. I've never seen batter being whipped so fast.

I looked at them, went out, then came back later. In their hurry to be elsewhere other than at the scene of the crime and around me, Sissy and the others had left the evidence behind. It was the Sunday newspaper comics. I went over and picked it up off the cabinet, and low and behold it was turned conspicuously to the comic strip **Li'l Abner.** It seems there was this ugly, feisty, simple, outspoken hillbilly woman named **Pansy Y**okum who had come into billions and bought land in Kentucky where she built this palatial estate, "Dogpatch," and was outraging "New York" society by her crude *lack of culture.* Did I see a resemblance between this and me?---I--I guess it wasn't a stretch.

But do what?! It couldn't be. This had to be some joke in bad taste, I thought. But no, no joke--it was a whole comic strip, now syndicated all over the U.S. and abroad with a following of millions. Who the hell was doing this? It said Al Capp, some unknown up until now, who was fast getting a reputation for the outrageous. I looked closely at some of the strips--biting, no-holds-

BLUE BLOOD

barred satire they called it, aimed at social hypocrisy and the upper crust.

One of the panels of Capp's comic strips that appeared during this time showed this "**Pansy Yo**kum" character going into the posh party of some high society woman. On seeing **Pansy Yo**kum, the slinky, sequence-studded socialite remarks to her equally haughty butler that this odd woman didn't look at all like a wealthy dowager to her. Just then Pansy helps herself to a seat, literally plopping down on the woman's nice sofa. The socialite approaches her and offers to take Pansy's hat, which is actually a country field bonnet. Pansy gives it to her and then sticks her cowboy-like boots up in the air practically into the socialite's face and says she'd just as well to take her boots, too, while she was at it. So uncultured and uncouth was **Pansy Yo**kum in this comic strip.

I may have not been that cultured, but I never told anyone to check my Texas cowboy boots along with my hat--and certainly not any farm bonnet--at any social gathering. But I could now see from the way I was being painted around Lexington by "high society" why the servants were sniggering along with the Thoroughbred people and others as they read about **Pansy Yo**kum in *Li'l Abner.* They assumed--or wanted to assume--that this character could only be a satirized me!

PASSIONS AND PREJUDICE

Had I ever been out of touch! I had been so busy that I had not noticed that the comic strip and I had been the talk of Lexington society and Kentucky for some time now in whispers and giggles among Kentuckians who were more savvy and up on "current events." Whether I was or I wasn't, everyone seemed to identify me with **Pansy Yo**kum. That little inside joke was enough to make many an old Kentucky society jewel gleeful. But there was nothing I could do about it. I was a public figure, and it was satire. In looking into it a little further, I found that Al Capp had quite a reputation for satirizing real-life personalities in his strip. It seems people could not figure out if he was a genius that deserved the Nobel Prize, an abomination that should be tarred and feathered and run out of town, or just a plainly dangerous man with a poison pen whose attention you didn't want to attract. Could this really be me in disguise--a very thin one at that? If so, was "Dogpatch" none other than my beloved Spindletop Hall and Spindletop Farms in Kentucky; and was New York society in Capp's comic strips none other than Lexington "high-horse" society satirized? Tell you the truth, that last part about possible satirization of Lexington society didn't bother me so much as the first part about me and Spindletop Hall. My Spindletop Hall and Spindletop Farms "Dogpatch?!" I suppose that could have been an allusion to the fact that I had built that dog room downstairs in Spindletop Hall for guests to put their dogs in, should they visit; and I had

BLUE BLOOD

furnished it with complete doggie amenities, including bath facilities and fire hydrants. Of course, it was no secret that I also raised Pomeranians, Scotties, and other types of dogs in various specialized kennels at Spindletop, including Llewellyn setters for field trials. I suppose, since I also had loads of other animals including birds, horses, and goats at Spindletop, I could be thankful that no one was calling Spindletop Hall and Spindletop Farms "Birdpatch, "Horsepatch, or----worse still----"Goatpatch."

Another panel in the comic strip that day showed the high society folks all up in arms and having an emergency town meeting since they thought this hillbilly Pansy Yokum was going to ruin their property values with her Dogpatch menagerie. I suppose that was a direct offshoot of that note that had been pinned to the front entrance gate of Spindletop when I first moved to Lexington asking me to please remove my Longhorn cattle from the front pasture since somebody thought they were scaring the Thoroughbred horses behind their well-kept white board fences along Iron Works Pike. At the time of that tacky request, I did move them to a back pasture. Now that I was reading this, however, I only wish I had left them all out there staring over the front pasture fence and taught every one of them to yell "BOO" instead of moo all the way across the street! Capp satirized many public figures and events of the times in his *Li'l Abner*.

PASSIONS AND PREJUDICE

Some said the author George Bernard Shaw was somewhere in the strip; members of the Dillinger gang were thinly veiled in it; the Gloria Vanderbilt custody case was there; women aviators resembling Amelia Earhart, and even actresses Jane Russell and Kim Novak weren't spared in later years. It seems no one was immune to Mr. Capp's sharp satirical pen--maybe I wasn't, either. My reputation had spread far and wide, I admit.

What a thought--that I may have been the inspiration behind Capp's hillbilly character, **Pansy Yo**kum, who invaded Kentucky with her "palace" and incensed the proper, rich folk. At times like these, a healthy sense of humor about oneself and others is about the only antidote. But, I admit, these suspicions did prick me in the side a little bit. But if people were busy talking about me, as the saying goes, maybe I was giving somebody else a rest. That was about the only positive point I could think of.

Cape and I had not given up on breeding Roxie Highland to Beau Peavine despite the heartbreak of the first foal dying of pneumonia. We tried again in 1938. The foal was due any day. Everyone was waiting, and there was always someone on watch at the barn with Roxie. Cape and I went out to her every day, checking and rechecking several times, often sitting with her in shifts for hours at a time. Roxie was always my weakness in horses. She was Mildred's horse, and

BLUE BLOOD

a world champion many times over. This would be some foal! The foal was coming on February 20th of 1939. We were all there--all the farm hands, the vet, Cape and I, for the proud moment. The foal was born healthy and rambunctious A few weeks after foaling, however, suddenly and unexpectedly, Roxie Highland died. We were able to find a foster mother for her foal; but Roxie's death was a tragedy that defied description for Cape and me. Cape was severely distressed; I was practically hysterical. Me, who always brought any hurt animal I found to Spindletop for care and often gave them a home when they had none. Me, who loved Roxie like she was my own child, like she was, in a way, Mildred, or all that was left of her growing up that I could see and feel within my reach at Spindletop. Through blinding tears, I felt the merciless pain once again of being forced to let go--again unexpectedly. It brought up all sorts of feelings that I thought time had somehow covered. We fool ourselves a lot in life--nothing is ever covered. It just lies there in the shadows patiently waiting for a nudge.

I buried Roxie Highland, perhaps the greatest three-gaited champion Saddlebred mare in history, and a horse that was more than a horse to me-- she was one of my oldest and best friends--in the beautiful little parkway just south of Spindletop Hall on the drive leading to the training stable. With her, we buried her show blanket, hood, halter, bandages, and other equipment. We covered

PASSIONS AND PREJUDICE

Roxie's grave with a blanket of lilies and other floral arrangements, and later erected atop the grave a granite monument topped with a bronze statue of this great mare that had become legend. No less would do.

Left behind was the five-week-old chestnut daughter of Roxie by Beau Peavine--a Spindletop breeding dream--my dream--that had turned into a nightmare when Roxie died so tragically. We named this alert and sprightly little offspring of two of the all time greats, Roxie Highland of Spindletop. What a beautiful foal she was! She would have done her mother proud.

We also had roadsters at Spindletop Farm. In 1940 we retired a horse at the Denver, Colorado, Saddlebred show that was proclaimed far and wide as the greatest road horse that ever lived. His name was Senator Crawford, and R. C. "Doc" Flanery of St. Charles, Illinois, showed him for Spindletop. There has been none like him since. He was so strong and powerful that he literally turned the wheels of the driver's cart out as he rounded the track to win championship after championship. We had to replace a lot of wheels that way; but it was the Senator's trademark on the track. What enormous power this great horse had! He was legend among road horses.

Beau Peavine made his last appearance and was retired at the Lexington Junior League Horse

BLUE BLOOD

Show in 1944. By then I had cooled down a little from the Junior League incident. Sooner or later, it's time to forgive and forget. I later donated the Beau Peavine Challenge Trophy in honor of Beau to the Junior League from Spindletop Farms, and it has been awarded every year since then to the best junior three-gaited horse in the Lexington Junior League Horse Show. The trophy remains in the possession of the Junior League of Lexington to this day.

I was lonely; it was like Spindletop Hall seemed to be getting bigger around me, and the servants fewer. And I was getting older.

Then guess who came knocking at my door-- my bedroom door. Cape Grant. He said, "Pansy, I love you. Pansy, I care for you. I have been divorced from my first wife. Pansy, will you marry me?"

I somehow wanted to hear those words. I wanted to hear them. I had been alone a long time--and lonely all too long. It was like I never got completely out of that desolate desert they called Orange, Texas, where I grew up among the ghosts and mirages that were lonely, dried-up, old people. It was like I was still being held there in some type of limbo.

Cape came in, and closed the bedroom door behind him. There was only the soft amber glow

PASSIONS AND PREJUDICE

of a distant night lamp on my dresser. As he stood in front of me that night in the velvet darkness tempered only by that soft glow in the background, so close was his body to mine that I felt the heat of his. He put his arm around me, pulling me close; and with his other hand he ran his fingers--haltingly--but with great control and assuredness of what he was doing--over my lips, then brushed the back of his hand over the sides of my face and around my neck and down into my breasts--as he kept my eyes held in his.----And there were feelings that I had not wanted to feel for so long aroused in me once more--feelings that I had buried so deeply since Frank's death. I had never wanted to feel again. It hurt to remember--and it hurt to feel. He began to unbutton my dress as he now kissed my face and breasts----and I gave myself over to those emotions of long ago--and I felt----I felt----after so long. And I began to kiss his hands as they passed teasingly across my face and down into my dress. He ran his other hand up my leg and onto my thigh, grabbing me forcefully in the crotch while kissing me; and there was a surge that went through my body as uncontrollable thoughts of the past began to flood into my memory, like a river whose dam had been weakened by the faintest of touches and whose waters could no longer be imprisoned-- a river whose waters were suddenly bursting forth out of control, flooding out all logic with a confusing blend of comforting emotions and primal instincts. It felt good--so good--to let go--to let go at long last--and to be a woman again--to relax down into a bed of feelings and emotions and touches and sounds and groans and sighs that drowned out the world of reason. I had been reasonable all too long. For just a little while, in my mind, in my make-believe world of stolen moments, in that flood of powerful emotions, I felt once again. I felt Frank's arms around me, his smell, Frank's lips on mine, his body--Frank's warmth surrounding me. It had been an eternity in exile; I welcomed him with my whole being. I was with Frank once again.

Cape and I made love that night. It was physical love, not real love--physical love that sells its stupidly willing victims of denial a bill of goods and then leaves them feeling a starvation and loneliness that dissipates into a nothingness and coldness and a worse void when it's over. It was the love being given by a younger man with a different agenda to a needy older woman whose defenses were down; a woman who dared to show her vulnerability because she was, for a few moments, just plain lonely, and tired--tired of fighting everything and everyone; a woman who wanted a brief respite of acceptance, warmth, and something reminiscent of love--even at the cost of its being a glossed-over counterfeit. Cape's "love" to me that night was animalistic in its brute thrust, heat, and fury, and as cold and calculating as any rape for power and control.

PASSIONS AND PREJUDICE

I woke up as a smothering heat and blinding sun barged ruthlessly through the window the next morning. As I opened my eyes from my self-induced dream, I found myself lying in the harsh white light of truth and the uncomplimentary sweat of reality. I looked at the man next to me. And I saw plainly. It was Cape Grant--not my beloved Frank.

On September 27th, 1949, Cape Grant and I married at Spindletop Hall. We took a cruise for our wedding trip--at my expense. Cape was twelve years younger than I was. I guess you could call it a marriage of convenience. Convenience for whom became the question, however. At first, I transferred a hell of a lot of stocks and stock rights into Cape's name. He liked that. He even said he loved me--more than once without noticeable feeling, and on appropriate occasions. He was, after all, a man of studied refinement, and proper form. But I seldom saw him anymore.----We lived in the same house;----we slept in separate bedrooms;----and he used my money.

Cape was a master at public show; and he was almost always solicitous of me around others. Like so many marriages, however, that look so ideal on the surface and to the public, it was entirely different behind closed doors. He was like two different people. Cape did not know it, but I would often watch him sneak out of the house at night, and from my bedroom window I

BLUE BLOOD

would see him get into his car and leave the grounds. He was having an affair. Rumor was, it was the wife of a Baptist deacon. I only know that I was just as alone as before--and worse--I was betrayed--and hurting again--and angry.

I decided to sell the horses. The horses were Cape's fame and claim to fortune. But for me, it was a relief from all the responsibilities I had been carrying. Only one hitch--I would always love those horses; in that respect it hurt me, too, maybe even more than Cape, since I never lost my conscience.

I remember the last race Cape and one of the only two Standardbred horses we had at Spindletop were in at the Red Mile in Lexington. I wanted to see his cocksure arrogance beaten, just for once. He was driving Hot to Trot, a fine chestnut mare with white markings out of the great young sire Hot Damn that was fast becoming legend. The particular horse that Hot To Trot was racing against had a real weakness for fresh green clover, and I knew it. Just before I left for the race, I loaded my purse full with green clover. When I got to the Red Mile, I took a close position by the finish line. Since I was the owner of the

PASSIONS AND PREJUDICE

horse that was favored to win, I could get right close behind the line.

Everyone knew Cape and the Spindletop Standardbred would win. Spindletop always won. And, sure 'nuf, here came Cape with Hot To Trot around the bend, pretty much nose to nose with "the green clover horse" who was the next favored. This was a heated race--and close, very close. Nose to nose they came with Cape looking confident. They were right upon the finish line when--oh, my, clumsy me--my purse just happened to fall open and all this fresh, fragrant, aromatic green clover just rolled out in front of the bulging eyes and flaring nostrils of the "green clover horse." Like jumping to its addiction, that horse literally leaped ahead of Cape and Hot To Trot over that finish line--by a nostril, and shall we say, two extended lips----and won. Cape never knew what hit him. (And if you think my purse just happened to pop open full of fresh-smelling clover at that critical moment, you were indeed born yesterday.) Cape didn't like to lose. He didn't take it well at all. I was, for once, mischievously delighted. All I did when Cape walked by was smile. Cape just glared and kept walking. I really don't know what got into me. Let's just say----the devil made me do it.

You would think that I would have at least felt bad about a Spindletop horse, a favored one at that, losing that day. But no. As I looked at Hot

BLUE BLOOD

To Trot taking second, it was almost as if I could swear that horse winked at me in her own horsy way. I think even she took it as a victory. I suppose you could say that we were on the same wavelength--in cahoots with each other. You see, that was the same horse that I found Cape jerking around in the barn one afternoon weeks before. I had walked in on one of Cape's training sessions. Before people, Cape appeared like a show horse's best friend. But Cape also had a rough side that people were seldom privy to. There were moments when, to get what he wanted, Cape could be downright mean to a horse without feeling a twitch of conscience. I know when I walked in the barn that afternoon, Cape had picked up a whip and was going at Hot To Trot and jerking her around by the bridle. The mare didn't like this and was putting up quite a fuss when I came in. Cape's words between clinched teeth to her were less than kind: "You son of a bitch. You'll do what I say or else." I was furious at what I saw. I came up from behind Cape, who was startled that he had an audience, and jerked the whip from Cape's hand.

"Don't you ever hit one of my horses!"

"Pansy--you bitch--get out of the way. Gimme that damned whip. I'll show this bitch who's boss."

"No! Leave her alone."

PASSIONS AND PREJUDICE

Still jerking the horse around by the bridle, "I'm going to show this bitch who's boss if its the last thing I do--now get out of my way." In his anger and determination to win out over the animal--and me--, he coldly pushed me hard out of the way, and I fell against the side of a stall. When he realized that he had shoved me, he stopped short, and was silent as he watched me regain my balance.

With perfect control, looking him dead-level in the eyes, I said as I stood, "I think she knows who's boss here, Cape--I am."

"Not with these horses, you're not." Then a sneer crossed Cape's face that reminded me all too much of the sneer that would cross Albert's face so long ago with that "what-are-you-going-to-do-about-it" attitude. It was as though I was having a flashback of that night years ago, when Albert had beaten me and thrown me out into the mud and rain of the night with that superior sneer across his filthy face. But this time, I was older. And I knew I had a choice.

Steadily, calmly, with an unmistakable fight behind it to back it up, I set the record straight: "I'm the boss here, Cape. ----Don't ever forget it."

Mockingly, "And what does my boss-lady wife think she's going to do about it?"

BLUE BLOOD

"I'm going to throw you square out on your ear if I ever see you doing anything like this again. And if you ever push me again"--I shook my head with more meaning than any words could ever convey.

Smiling that resentful controlling smirk and trying to gain control, "You know I'm the best trainer in this game. You wouldn't dare. Who'd you get to replace me? There's no one who can compete with me."

"I can, Cape--and I will if I have to. Don't press your luck." The smile dropped from his face. Cape, obviously enraged by the challenge, suddenly flung the reins down, turned abruptly, and walked briskly out of the barn. I picked up Hot To Trot's dangling reins. She was ears up, alertly looking at Cape as he exited the barn. To punctuate what had happened, the mare gave a deep, knowing, muzzled whinny, as if to say, "And put that in your pipe and smoke it, Cape." That's why, that day at the Red Mile, neither Hot To Trot being driven by Cape or myself minded coming in second--because it was a big win--for both of us ladies. The only one that really minded----was Cape.

To dig Cape and his superior attitude in the side a little more, I decided to indeed compete with Cape. Not in the horse arena. That would have been suicidal. But in his hobby arena. Cape

PASSIONS AND PREJUDICE

liked to show Llewellyn setters when he was not showing horses. He was a hunter and bought the best setters around to show off at dog shows. I decided two could do this. So I bought the best Llewellyns that money could buy in England, and brought them back to Spindletop. There was a show one Saturday where Cape was planning on showing his dogs. And I had entered mine right alongside his, just for the hell of it--a little "friendly" competition. Cape thought it was too funny, and he gave me a little advice on leaving for the show that day. "Forget it, Pansy, and don't embarrass yourself with dogs that can't win against mine." What a cocksure, cock-a-doodle-doo, I thought.

The dog show that day was spectacular. People came from all over the U. S. to show what their setters could do. Cape was on before me with his prized Llewellyn. Just like with horses, he was a perfectionist. And his dogs were both beautiful and just as perfectionistic. So I had brought some insurance with me that I hoped would even things out a bit. I had brought a dog whistle and had it stuffed down my dress between my breasts. Cape was doing quite well in the ring and was garnering admiring applause from the onlookers. That's when I decided to slip away from the crowd. I went behind a tree, brought out the dog whistle--so high pitched that only dogs could hear it, of course--drew in a deep breath, and blew it for all it was worth. Let me tell you,

BLUE BLOOD

every dog in the place went berserk--barking, howling, tearing away from their owners. Everyone's mouths dropped open. It was hard to say exactly what had happened, it was all so fast--a real mystery to everyone----but me. The people just knew it had never happened before. All hell had broken loose, and it looked like Cape's Llewellyn had started the commotion from inside the ring. Cape was in shock. It was his best dog. When things quieted down, and all the owners had their dogs back and under control, Cape was the only one fined points--and heavily fined at that--for lack of control, since his dog had been the only one in the competition ring. I was the next entry, so I signaled to have my Llewellyn setter brought in. I took him; and we walked into the ring with the dog at a perfect heel, and did our stuff. -- I made sure that the first place trophy I took home that day was especially shined to a sparkle and placed in a show spot on the mantel where the sun would hit it just right and give a gleam to it that would knock your eye out. It was for everyone's enjoyment, especially Cape's.

    There was a brief moment in here where I actually felt sorry for Cape. I've found over the years that there is a type of wrong pity that a person can have. Does a leopard change its spots? Funny how you can flirt with the most

PASSIONS AND PREJUDICE

unrealistic ideas, like "give him another chance, and he might." Again, does a leopard change its spots? Carve this one in stone: "No, they never do." I think I needed to write that on a blackboard over and over at least a thousand times, because I simply didn't have that one down.

All Cape and I seemed to do anymore was either avoid each other or fight. It was like there were no kindnesses between us--just acid words. Cape still drank on occasion. But that also seemed to have picked up again. After our first conversation about his drinking--that time when he and Owen were in that horrible wreck years ago, and then he let Owen go--for a long time after that it was like Cape was out to prove he was in charge of the alcohol instead of the alcohol being in charge of him. Yes, he was in complete control. But he was also in a con game of deception being run by a much superior adversary. When we had guests, he'd often say that he didn't want a drink because he didn't like to drink that much. He'd say, "Drinkin' is a full time job, ya know." A man like Cape could never admit to being out of control of himself or anything else. But no matter what he told others, he was becoming closer and closer friends with that little brown bottle, which was increasingly jealous of anyone who came between it and Cape. I felt the bottle would eventually suck him into it, and he would drown in that prison, never knowing what hit him. Inside Cape somewhere, he seemed to be embroiled in a battle royal that I was not completely privy to at this time; and his confidence of superiority over his adversary was dragging him further and further under. Cape had no problems. He seldom lost at anything. He never accepted anything but what he alone thought. And he was always capable, all by himself, to overcome anything and anyone. Cape not only needed no one, but he seemed to see no one but himself as the center of the universe. He was some proud, self-sufficient man. Whereas these characteristics might be viewed as good under some circumstances, where alcohol was concerned, they were deadly.

That's when I did one of the stupidest things I believe I have ever done in my life. In my mind I told myself I should actually practice what I preached--the Christian way. I should take it upon myself to mend some of this bitterness between us because Cape seemed to be either unable or unwilling to do so. If there was anything Cape excelled in, it was pride. So, I would be big about this for both of our sakes. There was something eating away at Cape that only he knew, and it was making me sick, too, I think, because for a few moments, I actually thought I could help him--help us or what was left of us. I sat down, swallowed my own pride, and tried to write him a note of reconciliation. I wrote it quickly without worrying about the wording. I wrote it from my heart, as honestly as I could, and as personally as I could without considering the dangers of opening

PASSIONS AND PREJUDICE

myself up so widely. I never understood how you could be both honest and guarded. The two never seemed compatible to me. The letter said:

> Dear Cape—
> I am sorry we can't get along. I think we both should do the best we can to try. That's all behind you as soon as a thing is brought out in the open it loses its force. Don't try to run away again from it. Stay here and whip it. You can do it with my help and God's blessings which you have got. I will do every thing for

BLUE BLOOD

you I can do. I know you will come out. You know two wrongs never made a right. Think what we have to lose and everything we have built up and all you have accomplished. Please excuse this letter. I will try and do everything I can, helping you. I will send this on so you will get this note.

With always
Patsy

PASSIONS AND PREJUDICE

I got one of the servants to take the letter up to Cape's suite.

Then I went into town. I wanted to get Cape something--something to help make things right again. I bought Cape a man's 18 karat white gold ring with six-diamonds. There were two small diamond baguettes on one side of the shank and three small marquis shaped diamonds on the other side of the shank. There was no real rhyme or reason to what I was doing. I told myself that when we discussed the note, I would give him the ring as a type of olive branch for a new beginning--if that were possible. Cape always liked expensive things. On one level, I suppose my actions seemed ridiculous and irrational even to myself; on another, I had actually talked myself into believing that maybe there was still some hope, and *I* could make things all right again. I think, at that point, I was simply sick--close to being as sick as I had been when I was under Albert's control so long ago. It was as if with my letter and with the ring that I were begging Cape for a few moments like an abused child not to hurt me again like Albert had so long ago. It was as though I were saying, "Please, please, if I am very, very good, you won't hurt me, will you?" It was strange how I went back to such sick behavior. What I was doing was totally against my will and totally out of my consciousness and certainly all wrapped in good reasons that I could swallow, like walking through an old familiar nightmare. I may have been kicking

BLUE BLOOD

and screaming, but the road I was sliding down was comforting in its familiarity. Something Cape was doing was reminding me of a twisted time in my life. It defied all reason. But when he did certain things around me, I began to react automatically with a behavior that I had allowed myself to be programmed with subconsciously long ago as a young girl by an abusive husband named Albert. I was like a sheep going to slaughter. Once again I was being enslaved by an invisible sickness that I believed was gone forever because Albert was no longer in my life. It was as though I were being held in a trance by a familiarity that was so effortless and so easy that it actually felt comfortable and right. But it was comforting me all the way to the slaughter block where my self-respect was going to be butchered right before my very eyes. And what was worse--what was worse was who was plunging the first knife into that self-respect. Surprisingly, it was I myself.

Cape did come by my suite that evening. He had my letter in his hand. I had the ring I had bought for him in my pocket and was turning it over nervously, over and over again, waiting for the right moment--waiting to see what he would say. I looked at him. I really don't know what I expected. What came out of his mouth that night, however, was as cold as ice: "I'm going to my house in Mississippi for a while; I have business there." Then he took my letter that he had been carrying in his hand, wadded it up before me,

PASSIONS AND PREJUDICE

looked at me, and then threw it in the trash can beside my door. He then left down the staircase and out the front door.

Not a word came out of my mouth during all of this. I stood at the top of the stairs for a long while, I think, only marginally conscious of the sound of Cape's car fading into the distance.

In the midst of a hideous thing happening, things seem unreal as the mind tells you it is suspending all judgments until it can make better sense of things, make things fit into some normal or recognized pattern again. You are in a mental limbo of sorts. If things become too unreal, it is though your body goes into a state of shock, and your mind goes into a type of protective blackout, during which you don't remember taking the steps you took or doing whatever you did. This happened to me that night. At least, the next thing I remember I was downstairs standing in the doorway to the library. I didn't remember ever coming downstairs.

Ed Fitzpatrick, who oversaw some of the operations at Spindletop Farm and who had been with me from the beginning, was there in the library that night, wrapping up his work on the books for the evening, getting ready to go home. He had served me and Spindletop Farms loyally for so many years. He lived with his wife in one of the tenant houses on the farm. He closed one of the books he had been working on on the desk. As it snapped to loudly, I suddenly awoke from my blackout and this nightmarish visit to the past, as if jerked by the scruff of the neck by a wolf mother teaching her pup a life-or-death lesson that must never be forgotten. I once more became conscious of my surroundings and of the ring that was still in my pocket, the ring that was to be my last try "to make things right" between Cape and me. In a way, that ring was my last link to the past. I suddenly saw what I had been doing as what it was--a remnant, a skeleton I was reacting to from long ago that was no longer welcome. After all these years that had passed, it was a ghost haunting me that I didn't even know was still around. We are more in the throes of our past and our experiences than we would like to believe--and the passage of time alone doesn't correct these things. When I realized what had been driving me toward this reconciliation with Cape--that I had been reliving some of those horrible times with Albert--it was a little like violently vomiting up something that had been making me feel all bad and nauseous inside. The secret of my bizarre behavior toward Cape----my writing the letter----my buying the ring----was suddenly out where I could see it clearly now and put it to rest, at long last. I knew right then and there that I would no longer accept invitations to visit hell. It wasn't a prime vacation spot. Deep in thought, I was still slowly turning the ring over and over in my pocket. It had been a very stupid idea

PASSIONS AND PREJUDICE

on my part. No one could help Cape but Cape. I would no longer be his "rescue ranger." He would now have to stand on his on feet. *I let go!*

I looked at Fitz--that's what I always called Ed Fitzpatrick, "Fitz"--and took the ring out of my pocket. "Fitz, I have something for you."

Fitz looked up, got up, and came over to me politely. "Well, Miss Pansy. I'm all done with everything for the evening."

"I want you to have this." I opened his hand and put the ring in it.

He looked at it wide-eyed, then at me: "Miss Pansy, I cain't take this ring."

"Please," I said, folding it up in his hand. "I appreciate all you've done for me here at the farm. I have no need whatsoever for this ring." His face was wadding up into one big question mark. Then, in a put-on cheerful voice, I tried to pick up the mood. I quickly unfolded his hand where he was holding the ring and pointed out a few things to Fitz. "And hey, this ain't no fake ring either, you can tell your friends. These here are two real diamonds, and these three are what they call marquis diamonds." I tried another quick smile in an attempt to seem more happy, folded it back in his hand, and ushered him to the door amidst his protests that he didn't see how he could accept it.

206

BLUE BLOOD

"You've more than earned it over the years, Fitz, more than earned it."

When Fitz left, I locked the door behind him, turned around, and leaned against it. The make-believe cheerfulness that I had put on dropped from my face like a heavy weight. I hurt. I hurt very, very deeply inside. It was like an almost incapacitating throbbing ache. Only then did tears begin to form in my eyes as I allowed myself to once again feel that pain. What a fool I had been with Cape. What an incredible fool. I remembered a question I had asked myself once: "Do leopards change their spots?" The answer broke through my memory like a storm breaking down any last resistance in its way: "No, they never do."

PASSIONS AND PREJUDICE

```
FINE CHINA                                              DIAMONDS
CRYSTAL         P. Edw. Villeminot                      WATCHES
SILVER             Lexington, Ky.                       JEWELRY
                 300 SOUTHLAND DRIVE 40503
```

JEWELRY APPRAISAL

Property of **Mr. William Fitzpatrick**
Address **1544 Heron Lane**
Lexington, Ky. 40503
Date **Sept. 25, 1987**

DESCRIPTION OF ARTICLE	Estimated Replacement Cost
Man's 18 Kt. white-gold six-diamond ring	
center diamond.. weight.. approximately 2.00 cts	
color......... G	
clarity....... SI.. estimated value of stone....	16,000.00
There are two small diamond baguettes	
on one side of the shank and three small	
marquis shaped diamonds on the other side	
of the shank,...estimated value of mounting......	1,800 00
Total	17,800 00
	(plus appli-
	cable tax)

FOR INSURANCE PURPOSES

These estimated replacement costs are based only on estimates of the quality of the stones (unless specifically stated that the stones were removed and graded). We assume no liability with respect to any action that may be taken on the basis of the appraisal.

Evelyn Caswell
Appraiser

COPY OF A 1987 APPRAISAL OF THE DIAMOND RING
THAT PANSY YOUNT BOUGHT FOR CAPE GRANT, BUT GAVE TO
HER OLD FRIEND "FITZ" AT SPINDLETOP FARMS IN LEXINGTON,
KENTUCKY. COPY COURTESY OF MRS. ED FITZPATRICK.

BLUE BLOOD

It was the beginning of the end. The Spindletop dispersal sale was held at Tattersalls in Lexington. There is something tinny and unreal about the voices of the auctioneers. It makes the auction somehow surreal, impersonal, heated, almost ununderstandable to the untrained ear, and almost inhuman, but ever so proper. Again I was wrestling with letting go. When our first horse, Miss Dixie Rebel, was led out, I felt nauseous. Cape had always said, "They will be as long seeing another Miss Dixie Rebel as they were seeing Roxie Highland. There is only one Miss Dixie Rebel, and she's never been defeated." It was as if all the sounds stopped--as if everything stopped, suspended in mid-air. It was as though my eyes had suddenly become like a telephoto lens on a camera and were suddenly zooming in tight, close-up fashion on the animal and its feelings in the ring. It was as though my feelings and those of the horse suddenly became enmeshed as one in that hall of strangers. We were engulfed in confusion--disoriented, the surreal confusion and disorientation, like an animal in shock, that comes when morals, values, feelings, business, practicality and being lost in it all uncontrollably crash together at top speed and pull at the sleeves of those with a soul, much like a helpless child wanting an explanation of some incomprehensible horror that is happening before him. It was almost as if my entire being wanted to rise to its feet and scream *stop*! The emotions were running so high in me that I stood up and left the auction with a sick

PASSIONS AND PREJUDICE

feeling in the pit of my stomach. It had all been fiction until the moment of its doing. Then suddenly reality--that painful reality of letting go-- grabbed me and realization set in. It was hard to let go of the horses that I had invested such a large piece of myself so completely and wholly in. I loved them so. The headlines read:

Miss Dixie Rebel Tops Dispersal at Spindletop

It struck me as particularly cold somehow: "$10,000," the newspapers said, "for Miss Dixie Rebel, four-year-old fine harness mare." That was all. The summation of an outstanding career by a horse that had given her all. All dollars and cents to the onlookers. Only not everything was a matter of dollars and cents. How could a lifetime of accomplishments, dedication, loyalty, and friendship possibly be measured in dollars and cents? I would miss her.

Beau Peavine and I had a type of spiritual understanding between us. Somehow between human and horse, the feelings and desires of both were translated in a transcendent way that we could both understand. I promised Beau Peavine that he wouldn't be sold but would be able to die

BLUE BLOOD

at the home he had known so long, Spindletop Farm. I truly believe he understood this and was thankful. Beau died a natural death at Spindletop in 1957. I was good for my word and had kept my promise to one of the all-time great Saddlebred champions in the world. Beau was twenty-eight years old.

After Beau Peavine died, things began to break apart at the seams. Spindletop looked as if it were growing around me, engulfing me like a mausoleum. My beautiful dream Spindletop! I was never accepted, no matter how much I did to be a part of Kentucky. And I was getting older. And I suppose when you get older, there comes a time when you just want to go home once more. I had had it carved in the mantel of Spindletop Hall when I built it, *"East, West, Home's Best."* And this was not my home no matter how hard I tried. *How could I have forgotten that?!* The last thing I wanted was to be buried in Kentucky. I thought to myself that would be the ultimate: not only not to be wanted by your neighbors in life, but not to be wanted by your graveside neighbors through all eternity. I wanted to go home--to Texas. Texas may not have been any better, but at least it was familiar and filled with memories of a happier time, and Mildred and the grandchildren were there. My relationship with Cape was nonexistent. I was unhappy. And I felt a type of urgency. I had already written Mildred a letter:

PASSIONS AND PREJUDICE

Dear Millie,

I am thinking of selling Spindletop Farms. Please pray that I do. And please pray that I do the right thing.

There was only one thing that I had to do before I left Kentucky. Cape and I, although still married, were seldom together--certainly not living together. One of his last gifts to me in our "marriage" was to enter a Spindletop horse, which we had kept back from the dispersal sale, in the Junior League Horse Show at the Red Mile in Lexington and then high and dry leave me and say, "You take care of it. I'm not riding Spindletop horses anymore." The horse was Chief of Spindletop, a true champion of champions.

There was no way on God's green earth that I was going to cancel that entry; and moreover, no way in hell I was going to let Cape get the best of me before the Junior League. Chief and I had started together years ago when Cape first took him out of the van in Beaumont as one of the cornerstone horses of Spindletop Stables; and we would end this thing together. He had proven himself over the years to be a true-blue, blue blood in every way and in every contest. His kind of "blue blood" was indeed worth gold. His supreme intelligence and grace and beauty in the ring was legend. No horse could hold a candle to him. I still remember how, when I first saw him there in Beaumont years ago, I looked into those deep knowing eyes and said, "From now on, boy, you'll be called Chief of Spindletop." And I remember how I thought this horse understood the meaning of what I said and was somehow proud of that christening. There was no way that I was going to withdraw such a champion from the contest ring and no way that I was going to let the name of Spindletop and Spindletop Stables be humiliated without a fight--and least of all no way that I was going to let this happen at the hands of Cape Grant because of default.

Over the years, I had become a good Saddlebred rider, having taken lessons from Cape himself in more ways than one. People could often see me riding my personal Saddlebred, Paris Grand, at Spindletop--and always with an entourage of what I called "goat people." We had a herd of Nubian goats at Spindletop that always amused me and followed me and my Saddlebred horse across the farm. To me, those goats always conjured up a funny picture in my mind when they stood in a group, the way their ears hung. They reminded me of some kind of highly intellectual university scholars at a graduation--with mortarboard caps on--curiously probing eyes, swishing their ears back and forth as they turned their heads from one side to another, as if conferring with each

PASSIONS AND PREJUDICE

other about some deeply important, philosophical thought--maybe from Aristotle. People who know goats know that look and know what I mean. Anyway, they would always follow me and my horse over the farm, which was a show in itself for people to see.

No one would ever be as good as Cape in the ring. I had no illusions about that. I had ridden in horse shows. But certainly no one would ever think I would be so gallish and stupid as to try to ride in the Junior League Horse Show at the Red Mile. It was fast becoming one of the top shows for Saddlebreds in the country, and the competition was stiff. *Quit* was not in my vocabulary. I might lose; but, if so, I would rather lose big and go down trying. We never know what we have in us until it comes down to the wire, and we pull out all the stops. Our potential, what we really are capable of, is a secret we all carry and flirt with finding out about from time to time. Some want to know the secret; others don't, simply because there's a high risk to finding this secret out. By discovering it, we risk once and for all putting all our grand notions about ourselves and all our grand dreams to rest by finding out what's really possible for us--and worse, what isn't and what never was. It involves a willingness to accept the truth about ourselves. That's frightening. It scared me to death. It's this one fear that keeps many of us in the world of dreams, I think, rather than really going for and

BLUE BLOOD

achieving our dreams. It's safer to "think" we can than to find out for sure. But to risk finding out that secret is also to know that we have really lived. In my case, I was being brought face to face with finding out about myself through no choice of my own. Some of us need a push to make that first dive off the diving board. Maybe it's a fear of water: we could drown.

The day of the contest came, and Cape made good on his threat; he was indeed gone. I would have given anything if he had been there. It would have given me a way out. I was scared, but I felt I had no choice. I packed up my riding gear and quietly had Ed Fitzpatrick take me over to the Red Mile in Lexington where the event was taking place. I went into a back cabin in the barn and dressed. I stationed Fitz outside the door as a lookout. He was to tell me when the horse in front of Chief was about finished. Time came. Everyone still expected Cape. But that day, I mounted one of the all time beauties and great champions of Spindletop Stables. I told Chief as I mounted, "Boy, I need your great heart behind me. I need you to perform a miracle today--one last one, boy--for me, and you--and Spindletop." We stood poised to go. Chief and I entered the ring alone--but with an impassioned determination, as if this were the moment for which we had been born. My only focus was the horse. Only on Chief and on his beautiful form. There was a gasp that went up from the stands as people looked at

PASSIONS AND PREJUDICE

their programs, and then at Chief and me, and then, somewhat stunned, they realized: my God--that's not Cape Grant up,--that's--why that's Pansy Grant! It was as if Chief knew in his soul exactly what I was attempting to do and exactly what he had to do for me. The love and respect we had for one another seemed to lift us out of the contest at hand. It was as though something outside of us both took over in that ring: there was a perfection and beauty and grace and concentration that commanded attention with its presence. The people in the stands were spellbound. It was as if that great horse and I were suddenly suspended in time. And it was as if he and I together blended into one in purpose and determination, and were transcendent in a type of slow motion where all the senses are ever so keenly aware--aware and sharpened and heightened way beyond both what would normally be humanly and "horsely" possible. That day at the Red Mile, he and I were making one last statement to the world--saying one last good-bye to those who would hardly miss us. It wasn't for them. It was for us--both of us--and for the name Spindletop and everything it had meant over the years. Chief--or a higher power greater than us both--God himself, I think--did perform the miracle for which I asked that day. And it was a true miracle, I'm convinced. When we left the ring, the crowd did not stand nor did it applaud. It just sat there peering, motionless, expressionless. The silence was stunning. The last thing I heard as we exited was the voice of the head event judge. It came over the loudspeakers as if echoing and resounding out of the halls of eternity to my ears, announcing: "Ladies and Gentlemen, the winners: Chief of Spindletop with Cape,--" (there was a clearing of his voice and correction--) "with--Miss Pansy Grant, up." Chief and I never looked back. But suddenly as I rode out I saw what must have been a vision or some kind of premonition--a vision of what was to come, just maybe not in my lifetime, or in the lifetime of Chief. It was as though we were transported into some time in the future, and I saw the crowd at the Red Mile again-- and this time, things were different--this time the crowd stood--Kentucky stood--people who had come from all over stood--and slowly, ever so slowly, as in that same slow motion gait that I had experienced in my mind that day with Chief in the ring--they began at long last----to accept----to accept and to applaud----and this time, it was the applause of respect. Then the vision or daydream was over, just as quickly as it had come, and the reality was there before me once again. I looked at Ed Fitzpatrick who was waiting for me outside the ring. I dismounted, once more looked at Chief, who seemed to know what he had done and was proud. I rubbed his great head once more, and pressed my face against his, enveloping my arms about his long, graceful neck as far as I could, in one last attempt to convey to one of the most magnificent champions in history what he had meant to me over all these years. He knew. I'm

PASSIONS AND PREJUDICE

sure at that moment he knew. I then gave the reins to Fitz-----and I went home--home--this time, to Texas.

I bought the old Rothwell home at 124 E. Caldwood in Beaumont. We referred to it as the Caldwood Estate. Mildred and Ed, her husband, had moved from "El Ocaso," our house on Calder, to a beautiful home at 650 Thomas Road by that time. And I made a decision to have the house at 1376 Calder that Frank and I and Mildred had spent such happy days in together, "El Ocaso," which many called "the Frank Yount home," torn down. This met with a lot of opposition from Beaumont architects and historians. But my reasoning was that I was getting older, and you just don't know what will be done with a place when you're gone. "El Ocaso" was special to Frank and me. It was indeed the Frank Yount home: our very first home in Beaumont after Spindletop came in; and we were happy there, so happy, during Frank's lifetime. Mildred had grown up in the Calder home, and a lot of her private tutoring had been done there; and Frank--the only man I ever really loved-- had died in our Calder home on the eighth anniversary of Spindletop's coming in; we had even had Frank's funeral service there. I didn't want to take the chance of that house being used for anything other than a home, and certainly not just a home to anyone. Some people even tried to go around me and appeal to Mildred, as if I didn't know what I was doing. But I did know.

I had that home demolished. That place was like hallowed ground to me, and I wanted no one desecrating the memories stored inside its walls.

Before I left Kentucky, I had given Fred Wachs of Lexington, certainly my closest Kentucky friend and confidant over the years and owner of the *Lexington-Herald* newspaper, power of attorney to sell Spindletop Farms. When I got the farm, only my name appeared on the deed. Now that Cape Grant was officially my husband, however, I had to get Cape to sign power of attorney on the property over to Fred Wachs to sell Spindletop to the University of Kentucky. Cape was almost always gone. But I caught up with him in Booneville, Mississippi, at his Prentiss Court home, on one of his "business trips." I got him to sign the power of attorney over to Fred. His reticence was ominous. We had no relationship. It was like I didn't know this man and never had. You never know anyone until you get down to money.

I had originally approached the Catholic Diocese of Covington in Northern Kentucky about Spindletop. I was, after all, Catholic, and in their case I wanted to give it to them outright. But the Bishop thought it was too far to Lexington.

PASSIONS AND PREJUDICE

Catholics! Of course, I should talk. I was one of them.

I then told Fred to get in touch with the University of Kentucky which had already been in contact with me through one of my few friends, then--Governor A. B. "Happy" Chandler. Two things, I think, made and kept Happy Chandler governor of Kentucky for so long and one of the most popular politicians in the state's history: one, he could remember everyone that he ever met and call them by their first names; and, two, he could sing, "My Old Kentucky Home." We had an ongoing joke. I told Happy when I heard him sing "My Old Kentucky Home," I felt like breaking out in my own rendition of "The Eyes of Texas Are upon You." To avoid that, he usually refrained from singing in my presence. I had met Happy as a young man when he was campaigning for office, and I had liked him immediately. He then invited me to the very first Kentucky Derby Breakfast ever given in Kentucky before the Derby. It was hosted in 1936 by him and his wife Mildred, who he fondly called "Mama." Happy always said, "Surely goodness and mercy shall follow us all the days of our lives, and we shall dwell in the house of the Lord forever." That was his favorite saying. I always told him that I hoped that's where we both landed up. Then we'd laugh; and I'd say in good humor, "You never know, Happy. We're both pretty ornery."

BLUE BLOOD

Dr. Frank Peterson, one of the best finance people the University of Kentucky and the State of Kentucky had, had also asked Fred Wachs to talk to me about getting Spindletop for U. K. Dr. Frank Dickey, President of the University of Kentucky, then came down himself to Beaumont and talked to me personally about an arrangement with U. K. I was just going to give it to U. K. and Dickey. But Edward, Mildred's husband, was there the day he came; and Edward was always the attorney. He discouraged my giving it outright to U.K., and I excused myself from Dr. Dickey--we had been sitting on the front porch--and went inside for a private family "chat" with Ed. So, for what amounted to family peace, I compromised and ended up letting it go for $850,000, gift purchase, the price of the fence around Spindletop Farms. I wanted someone to have Spindletop that would really appreciate it and enjoy it. And I wanted people to at last be able to enjoy it, one and all. I wanted no one excluded from Spindletop. And I thought U. K., as a Kentucky institution of higher learning, would honor my wishes and appreciate it as the gift it was, and people would get some use out of it. One thing I sure didn't want was for it to go to any of the Thoroughbred horse people or horse farms adjacent to Spindletop.

Selling Spindletop broke my heart. I had loved Spindletop Hall and the beauty and grandeur of Kentucky from the beginning: they were my passions for so long, and a dream that Frank and I

PASSIONS AND PREJUDICE

had dreamed together, come true. I remember pausing before I put my hand to the deed to Spindletop for U. K.; and, almost as if I went into some daydream, it was like I was back in the halls of Spindletop in Lexington. It was as though I was looking at it from the inside, one last time. I ran my hands over the mantel of the oak room once more, with my fingers reading in the relief, "*East, West, Home's Best.*" As I stood there in those mighty halls, I turned slowly in place, taking in a magnified version of everything one last time; and I remember thinking to myself, "I suppose, when you build something like this up----and you put your whole self in it----your whole heart and soul---that sooner or later----you have to let it go!" And it was as though I was looking at Frank's portrait once more at Spindletop and talking to him once again: "I let you go once, Frank."

It was times like these when I wished we were back on that Hill at Spindletop. It was simpler then: me, with my second-hand lantern; Frank, with that white linen suit of his out there under that big Texas moon. But it wasn't Spindletop Hill anymore----and it never would be again. Those were only shadows now, secrets hidden in time.

Signing the deed to Spindletop over to the University of Kentucky was not easy. Spindletop was a part of me, and I was a part of Spindletop. And so it would always be--through all eternity. But I signed. And I grieved a grief that never really left me. Happy Chandler proudly made the first announcement of the acquisition for U. K. before the Optimist Club in Lexington.

This final move on my part upset the rest of Cape's applecart, but good. As still my legal husband, he had been sitting pretty until then; but he could see the writing on the wall when I actually went through with selling Spindletop--which I don't believe Cape ever really thought I would do--and when I also faced him simultaneously with his infidelity, which he thought would never happen since I couldn't possibly know about his dirty little secret. I told him I knew about the affair he had been having all these years. Since she was reportedly the wife of a Baptist deacon, I asked him if they read scripture while they did it or if his choice of partners made the act of adultery a little more righteous in his eyes. He slapped me. I told him to get out. He paused in the doorway and turned to me, "She wasn't the only one, Pansy." I picked up a lamp and shattered it on the door as he pulled it to.

Two weeks after I sold Spindletop to the University of Kentucky, I think in an attempt to save face after being caught with his pants down and in an attempt to beat me to the draw, guess who came knocking at my door again----Cape Grant. He said, "Pansy--I don't love you. I have never loved you. Pansy, who could love someone as ugly as you are!? Pansy--I'm filing for divorce.

PASSIONS AND PREJUDICE

And I want to remarry my first wife. And, by the way, I'm going to take you for every penny you have!" I told Cape in no uncertain terms to get out and never come back, and it would be a cold day in hell before he got one cent of my money!! And I also told him it was always Frank Yount who had loved me, and I had always loved Frank. I was Pansy Yount, not Pansy Grant, and I always had been----and I would always be. Cape just laughed. Then he left.

The next day, March 3, 1959, I was served with divorce papers and a restraining order, which was a joke, bright and early; Cape had been all prepared, it seems, for some time now. The papers said that Cape had been the brunt of undue cruelty at my hands, that I had hit him, and that I had been neglecting my wifely duties to him. He went on to say that I had a violent temper, and at one time had tried to shoot him, but luckily he had been able to disarm me.

Hogwash! No, I had not tried to shoot Cape. It had unfortunately never occurred to me. And secondly, if I had taken a pot shot at him, he would have been a hen and not a rooster; I was a crackerjack shot.

It seemed that Cape apparently thought this was some kind of *fait accompli*. The truth was, he was never more wrong. Every inch of me was a fighter. If he wanted to play real hardball, I was game! I felt one of those "now-the-fur-is-really-going-to-fly fights" coming on. And I was determined it was not going to be my fur. So on March 12, 1959, nine days after Cape filed suit, I filed an answer to his complaint and a cross-action suit against Cape. I can only say I was lucky to have gotten Spindletop Hall out from under Cape in time and into the hands of the University of Kentucky before the whole roof fell in and everything came to a head. Otherwise, U. K. may not have gotten Spindletop Hall and Spindletop Farms at all, and it may have been tied up and endangered in court proceedings for some time to come. And who knows what history may have written then? But Spindletop Hall and Spindletop Farms in Kentucky were now out of Cape's reach, as of two weeks ago.

In my answer to Cape's suit, I vehemently denied each and every one of his accusations, and I countersued for divorce *from him* on the grounds of "harsh and cruel treatment and conduct" toward me. And furthermore, I asked the court for my name back, that is, that it be formally and legally changed back to Pansy Merritt Yount. I also asked the court to adjudge that not a dime of my money belonged to Cape in any way.

This was the first time in my life that I had no difficulty whatsoever letting go. The whole world seemed to be following this divorce case through the newspapers, and garnishing it with public

PASSIONS AND PREJUDICE

gossip. It was the talk of the times. And I was determined that when this thing was over, the world could stick a fork in Cape, because *he would be done*!!

Then on December 17, 1959, the court handed down its decree concerning the suit and countersuit claims with these respective findings by the jury after three months of a knock-down, drag-out showdown of nerves in court, and a battle royal before the public:

SPECIAL ISSUE NO. 1

Do you find from a preponderance of the evidence that the acts of the defendant, Pansy M. Grant, toward the Plaintiff, W. C. Grant, if any, constitute such excesses, cruel treatment or outrages of such a nature as to render their further living together as husband and wife insupportable?

You will answer "We do" or "we do not."

Answer: *"We do not."*

BLUE BLOOD

<u>*SPECIAL ISSUE NO. 2*</u>

Do you find from a preponderance of the evidence that the acts or conduct, if any, of W.C. Grant toward Pansy M. Grant constitutes such excesses, cruel treatment or outrages of such a nature as to render their further living together as husband and wife insupportable as that term is defined in this charge?

You will answer "We do" or "We do not."

Answer: "We do."

Short and sweet, what this meant was that the court found for me, and I was completely vindicated from Cape's accusations. The court granted me a divorce from Cape, and *not* vice versa. And I got my name back: I was once again--Pansy Yount. And Cape did not get a dime.

Let me tell you, Cape was mad; oh, he was blistered! *So he appealed the damned thing!* And we had to go through the whole thing again, with all the accusations, with all the publicity! But then, the day before the court's decree was to be handed down----*the judge died!* Our court files were already over six hundred pages of testimony and motions alone--that's at least knee-high!--And do you know what? We had to go through the whole thing again!!! People had it practically

PASSIONS AND PREJUDICE

memorized by now, and the papers were happy to have something to feed on once more. But this time, the day before the court's decree was to be finally handed down----*Cape died!*

But I can't say, from my side, that I was sorry. I was not sorry.

Cape had been in Dallas during this time, where he had a home. He had gone down to where Nola and his family were. And out of nowhere he was struck with a stroke, "malignant hypertension," due to polyceythemia due to arteriosclerosis. He was immediately put into the hospital there. He was in the hospital for four days. Two hours before he died, he remarried his first wife, Nola, who was by his side.

Someone told me that Cape Grant's last words to his sons were, "Never leave the mother of your children." He then died. Cape Grant, one of the all-time greatest Saddlebred trainers and riders in history, the likes of whose natural talents, savvy for the show ring, colorfulness and flashy style will never be seen again by the Saddlebred industry, died May 28, 1962, at the age of 63, and was buried in Restland Cemetery in Dallas, Texas.

Cape and I had literally spent the last part of our lives embroiled in a knock-down, drag-out divorce for nothing but "stuff."

BLUE BLOOD

Now, I find myself looking back over my life; and three observations stand out in relief after all these years: (1) I am astounded at the brevity of life; we are here for only one brief, precious moment, and then, it's over; (2) We're much more complex and unpredictable and checkered than we would like to think: like the old adage says *"there's a little good in the worst of us, and a little bad in the best of us;"* and (3) I am convinced that life is not about getting or grabbing or accumulating or achieving, but about *letting go*--a foreign term to human nature and will, with the final letting go being death.

The time came--and I let go--this time, I would like to think with some grace and acceptance that it was simply my time. And I remembered in those last moments a conversation I had had with an old friend years ago in Kentucky, and it brought a faint smile to my lips once more:

"Surely goodness and mercy shall follow us all the days of our lives, and we shall dwell in the house of the Lord forever, Pansy."

"I hope that's where we both end up, Happy. I don't know though. We're both pretty ornery."

EPILOGUE

EPILOGUE

Pansy Yount died at 11:30 a.m. Sunday, October 14, 1962, at her Caldwood Estate (now called the Steinhagen Mansion) at 124 E. Caldwood in Beaumont, Texas, of kidney failure, only four and one-half months after Cape Grant died of a stroke in Hillsboro. Requiem Mass was said for her by Monsignor E.A. Holub, pastor of St. Anne's Catholic Church, where she was a member. Her obituary was simple and unpretentious. The University of Texas and the Masonic Home for Children in Lexington benefited from her will, as did a close friend, and one who served her faithfully for many years. Only two came down from Kentucky to attend her funeral in Beaumont: her longtime, trusted friend, Fred Wachs, Sr., who had stood by her through the years, had helped her give her beloved Spindletop Hall and Spindletop Farms to the University of Kentucky, and had testified on her behalf in her divorce from Cape Grant; the second one who came was Ed Fitzpatrick, "Fitz" as she called him, the farm hand to whom she had given the ring originally intended for Cape Grant in her failed reconciliation attempt. These two now served as her pallbearers.

Then came the auction of her personal belongings. As one person, who had known Pansy Yount personally, walked through the disarray in

PASSIONS AND PREJUDICE

the auction pavilion, she paused, according to newspaper reports, beside a large table-sized music box that had belonged to Pansy. She opened it, and the music box began to play *Auld Lang Syne.* The melody could be heard resounding *hauntingly* throughout the auction hall that day.

Could it be that Pansy's ghost still haunts her beloved Spindletop Hall in Lexington? Some say it does. Stories abound, as do sightings. On quiet nights, very late or in the early morning hours, workers and security guards tell of seeing what looks like a woman dressed in black in the hallways of the second floor.

Too, was Pansy trying to whisper to us through the words and haunting melody of *Auld Lang Syne* one last important secret wisdom from beyond the grave that day as she let go for one more time in that auction house?

Should auld acquaintance be forgot,
And never brought to mind?
Should auld acquaintance be forgot,
and days of Auld Lang Syne?
For Auld Lang Syne, my dear,
For Auld Lang Syne--
We'll take a cup of kindness yet
For Auld Lang Syne.

EPILOGUE

On January 30, 1969, Pansy's daughter, Mildred, died of an apparent brain aneurysm at age forty-eight while in Manitou, Colorado, closing on their summer home, Rockledge, where she and Ed Manion had first met and had fallen wildly in love. Her death was totally unexpected.

Ripples of Spindletop's accomplishments are still being felt today. In 1988 Spindletop's premier stallion, Beau Peavine, appeared twentieth on the national list of stallions in order of his impact on the nation's Saddlebred foal crop. Kentuckians who have never heard the complete name *Beau Peavine* or of Spindletop itself still can be heard saying, even in the tucked-away, back hills of Kentucky, "If I could just get a horse with just a drop of that Peavine blood in 'em, I'd outshine all of them other horses and riders." So legendary is the name *Beau Peavine*, even today! Such people may not know who he was, but they know he was "damned good at something." And they'd like to buy into a part of that greatness and celebrity.

SPINDLETOP NOW

SPINDLETOP NOW

It was Pansy Yount's passion for Saddlebreds that gave rise to the first Saddlebred Horse Museum, which was first located in a carriage house at Spindletop Farms itself. The museum has since been relocated and is now the American Saddle Horse Museum at the beautiful Kentucky Horse Park, located at 4093 Iron Works Pike, Lexington, just down the road from Spindletop Hall. It hosts thousands upon thousands of visitors from all over the world each year--a must-see for tourists.

Pansy's unique and picturesque carriage collection--second in the U.S. only to that of the Folger Collection--has been moved from Spindletop to the lower level of the International Museum of the Horse, also located at the Kentucky Horse Park, where it gives visitors a glimpse into an opulent and romantic past.

Likewise, the bronze statue of Roxie Highland, Pansy's beloved champion three-gaited mare that used to stand proudly behind Spindletop Hall is no longer there, but has been moved to the Kentucky Horse Park for viewing. The actual grave of Roxie Highland, however, is still located in its

PASSIONS AND PREJUDICE

original place of honor in back of Spindletop Hall, with a bronze cover that is inscribed.

Pansy Yount did the University of Kentucky a favor by giving Spindletop to them as a gift purchase. Not only was it quality land and inordinately cheap to acquire; but, even today, according to Dr. Frank Dickey, past President of the University of Kentucky, when U. K. is trying to recruit quality faculty to come to U.K., if all other considerations are equal between U. K. and another institution that they are considering joining, then Spindletop is quite often the decisive factor in that quality faculty member choosing the University of Kentucky over another institution.

Spindletop Farms in Lexington may no longer be the 1,000 acre plus grand spread it once was; indeed, over the years, it has been sized down to fifty-two acres. Too, Pansy's beautiful gardens that she cherished and hoped that others would enjoy, modeled after those at the Palace of Versailles, that once lavishly covered the grounds before Spindletop Hall and behind Spindletop Hall in the form of sunken rose gardens, have long been abandoned and are no more. Yet Spindletop Hall in Lexington, Kentucky, is still *uncommonly beautiful* and downright amazing and *haunting* in the way it tenaciously hangs in there--much like Pansy herself in life. In that respect, something of Pansy's determined nature must still be an integral part of Spindletop Farm's structures and grounds.

SPINDLETOP NOW

The mystery, rumors, gossip, and secrets that surround Spindletop still cast an alluring, hypnotic spell on those that wonder about its past and its lady to this very day.

In Texas, there is also the ongoing dispute over the Spindletop fortune from the oil well in Beaumont, where still today the Jefferson County Court House in Beaumont receives hundreds of letters per month from people claiming to be heirs to a piece of Spindletop oil money. These queries date back to the years of the first Spindletop oil gusher, as well as span over the last thirty-five years since Pansy Yount's death, and they number in the thousands. They have touched off a wild streak of complicated law suits in Texas over the years in hot pursuit of Spindletop money. Still on docket at the end of "*1996*" in the U. S. District Court in Beaumont, Texas, was a claim to any piece of Spindletop "profit, proceed, royalty, gain, advantage, title, or interest" that dates all the way back to a 1911 property deed involving people who believe themselves to be Spindletop heirs. Named as defendants in this suit are a slew of oil companies, as well as individuals. The defending oil companies in this suit, not including individuals named as defendants, comprise quite a distinguished who's who in the oil industry. They include the following oil giants:

PASSIONS AND PREJUDICE

Chevron, U.S.A., Inc., a California corporation, successor in interest of the Gulf Oil Corporation;

Amoco Production Co., a Delaware corporation;

Mobil Oil Corporation, a New York corporation;

Phillips Petroleum, a Delaware corporation;

Sun Oil Company, a Pennsylvania corporation;

Oryx Energy Company, Dallas, Texas;

Pure Oil Company, a subsidiary of Union Oil Company;

J. M Guffey Petroleum Company, a Texas corporation;

Lancier Oil and Gas Co., of the state of Texas;

Texas Gulf Inc., of the state of Texas;

Gulf Oil Corporation, of the state of Texas;

Exxon Gas Systems, Inc., of the state of Texas;

Gulf Refining Co., of the state of Texas;

SPINDLETOP NOW

Amoco Oil Company, a corporation of the state of Texas;

Karburhn Oil Company, a corporation of the state of Texas.

It is reported that two FBI agents who went to investigate Spindletop claims concerning money possibly due to those staking claims to being Spindletop heirs apparently disappeared without a trace some years ago. Thus the mysteries surrounding Spindletop continue. It is rumored that if all proven heirs and "heirs in waiting" to "rights, profits, proceeds, royalties, gains, advantages, titles, or interests" in Spindletop, who believe they are entitled to money were to be paid, it would break the State of Texas. Although thirty-five years have passed since Pansy Yount's death, Spindletop is still embroiled to this very day in controversy, intrigues, and possible "hush-ups." Thus, the haunting question: What dark secrets of Spindletop, like specters unrelentingly trying to break some mysterious, supernatural veil, are still waiting in the wings to be brought to light?

To date, little attention has been given to marking the way to Spindletop Hall's location near Lexington, Kentucky, with historical markers or signs. It would seem that it deserves at least some signs up along the way for Kentuckians and

PASSIONS AND PREJUDICE

tourists alike, if for no other reason than its rich historical significance that touched the world indelibly. It is easily the best kept *secret* of the Bluegrass, which could bring a wealth of tourism dollars to the Bluegrass along with the likes of the Kentucky Horse Park and the American Saddle Horse Museum, all on Iron Works Pike only seven miles from Lexington. It stands to reason that at least some of the dollars from such a table might fall into the lap of the University of Kentucky (and Spindletop Hall itself, which is self-supporting)-- which probably wouldn't hurt either and perhaps would even increase U. K.'s prestige as an institution of higher learning, which, as an educational leader, certainly values the importance of history.

When I first performed my one-woman show on *Pansy Yount of Spindletop* at Spindletop in 1988, I was struck by the fact that there was a picture of Miles Frank Yount in the entrance foyer, but only one small and uncomplimentary charcoal sketch of Pansy Yount on the mantel of Mildred's Suite upstairs. You saw it only if you looked real hard, and then you might miss it. I saw it, and then I didn't see it. It seems that someone had taken it down and put it in a musty closet on the second floor of Spindletop because, as one of the people instrumental in securing Spindletop for U. K. said, "They didn't think she looked representative enough to be in Spindletop." I found the sketch in a closet. I was shocked, and, therefore, without

SPINDLETOP NOW

thinking too much, worked this into my performance for the Board of Directors that night. There was a noticeable gasp that went up from the audience when I said, as the character of Pansy Yount that night at Spindletop, the following:

>**Pansy:** *(Picking up photograph of herself from nearby table; to herself, almost inaudibly and deep in thought:)*
>
>"Pansy, you're really not all that pretty, you know. Someday, after you give this to U.K., maybe someone will come along and take this picture of you and throw it in some musty closet somewhere because they won't think you're representative enough for Spindletop Hall; or maybe--just maybe--some dear soul will take it, and touch it up, and give it its proper place in Spindletop Hall. How I have loved you, Spindletop!"
>
>*(Puts the picture back on the table.)*

Not too long after that, someone from Spindletop who had seen me that night came up to

PASSIONS AND PREJUDICE

me at another performance I was giving for the University of Kentucky Donovan Scholars, and said that he thought I would like to know that "they" had taken the sketch and were touching it up as an oil, and making Pansy "look as good as we can" and were planning to put it up in Spindletop Hall.

About a year later, when I was performing for a group at Spindletop, there was a new addition to the second floor--an almost full, life-sized portrait of Pansy in oil that had been recently put up. It was placed at the top of the stairs in the hallway on the second floor and lit nicely.

There are many good-hearted and generous people in Kentucky. But there is only one thing still lacking in the portrait matter, as other people after performances who realize the real story of Spindletop have also commented on and been concerned about: perhaps Pansy Yount's picture belongs on the first floor of Spindletop in the foyer where it actually once did hang when she lived there (where now the portrait of Miles Frank Yount hangs); and Miles Frank Yount's portrait might take the place of Pansy's on the second floor, since Spindletop was historically Pansy Yount's creation: Frank Yount died in Texas long before Spindletop Hall was constructed and before Pansy came to Kentucky. It's a matter of giving credit where credit is due.

SPINDLETOP NOW

Still another part of that performance that night at Spindletop were these reflections from the heart of the character:

Pansy: *(Thoughtfully, looking out the window at her beautiful gardens she had modeled after those at the Palace of Versailles and taken such loving care of over the years:)*

"And will my beautiful gardens always be here for people to enjoy; or after I am gone from here, will they be ravaged, and raped, and left barren like the Hill at Spindletop so long ago?"

(Running her hand over some precious antiques and furnishings she will leave at Spindletop, and stopping before the two Stradivari violins, one inscribed with the nameplate "Mildred Yount," the other with "Frank Yount" in a glass music case in the music room of Spindletop Hall:)

"Or will pieces of my furniture or these Stradivari violins make their way over the years into living rooms that would not have wanted the likes of me to darken their doorways."

(Looking around the room, longingly:)

PASSIONS AND PREJUDICE

> "Or will it be that some dear soul will come along and restore you to the splendor of the day that was once yours? --May God keep you,--my precious Spindletop! And may God bless us all."

Indeed, today Pansy's gardens are no more; and the violins of Frank and Mildred that were once in the music case of the music room, which I saw on my first trip to Spindletop Hall during the filming of the movie *Sylvester,* are no longer in that case for visitors to enjoy.

It's as though a child had built a beautiful sand castle on the beach of a great ocean which enhanced the ocean's beauty, and then the ocean tides came in and erased it so no one ever knew it had been there. Pansy Yount and Spindletop Hall and its true history and heritage have been so obscured by the tides of time. And in Kentucky, Spindletop Hall and its lady have been reduced to a large extent to a building without a history and a building without a human face behind it.

Faculty members and their guests from audiences after a performance have come to me and said, "Spindletop is alive to me for the first time. We've got a treasure here in the Bluegrass, and it shouldn't be buried!" Visitors to conventions from out of state who have witnessed performances of Pansy Yount's life at

SPINDLETOP NOW

Spindletop have identified heavily with her as "Everywoman." I understand that now even tour buses from the American Saddle Horse Museum slow up a little as they pass by Spindletop Farm and Spindletop Hall, after one of its National Tours saw a performance of her life at Spindletop and were moved by this woman--her struggles, wherewithal, and compassion.

Spindletop Hall is more than a magnificent mansion in Kentucky, or an oil field in Texas that changed the world, or a stables that left an indelible mark on the nation's Saddlebred industry: it is the heart and soul of a person--the universality and humanness of whose personal struggles speak to the souls of people everywhere, and to people of all ages, and to people from all walks of life--because it mirrors in large part our own personal story as human beings.

Was that maybe what the music box with its haunting melody was trying to tell us that day in Beaumont when Pansy's estate was being auctioned? Are we indeed only as sick as the secrets we keep, both as individuals and collectively as members of the human community? And was Pansy perhaps offering, with the eternal wisdom of the ages now at her disposal, and in the kindest, most unobtrusive way possible, a word of reconciliation to her beloved adopted state of Kentucky and to anyone else who might

PASSIONS AND PREJUDICE

listen? Was Pansy Yount offering a word of reconciliation to anyone who may have been offended by her "foreign" Texas manner, or to anyone who might themselves feel guilty, or feel they might be some type of unwilling partners in a collective guilt that Spindletop Hall's very existence in Lexington seems, at least, in their minds, to accuse them of, so much so that they wish that Pansy Yount's story--and thus the history of Spindletop Hall and Spindletop Farms--would simply just go away and be buried once and for all? Was Pansy Yount, in all compassion, trying to reach out and say to such as these that day at the auction of her worldly belongings: **"It's time to let go? A new time has come. The past doesn't have to be hidden or buried or kept secret anymore because the secrets are no more. And we are now free from that burden, at long last--free to remember with kindness and respect and acceptance--the days of Auld Lang Syne.**

Should auld acquaintance be forgot,
And never brought to mind?
Should old acquaintance be forgot,
and days of Auld Lang Syne?
For Auld Lang Syne, my dear,
For Auld Lang Syne--
We'll take a cup of kindness yet
For Auld Lang Syne.

But *wait.* Maybe we should take a closer look still. Is there, perhaps, *one more*

SPINDLETOP NOW

secret, a message from beyond the grave from one who must be regarded, from the way she handled the obstacles that arose in her life, as noble and "true-blue"? And, if so, is it the message being whispered from deep within the little-known last stanza of *Auld Lang Syne*? That stanza reads:

And here's a hand, my trusted friend,
And give a hand of thine;
We'll meet again some other night,
for Auld Lang Syne.

For Auld Lang Syne, my dear,
For Auld Lang Syne--
We'll meet again some other night
For Auld Lang Syne.

APPENDIX I

A Dollar Bought a Lot in 1930

Champion Beau Peavine as painted by George Ford Morris in 1933; from the collection of the American Saddle Horse Museum, Lexington, Kentucky.

The Beau Peavine Perpetual Trophy. Pansy Yount supported the Lexington Junior League Horse Show. She gave them, in honor of Spindletop Farms and the legendary champion Beau Peavine, the Beau Peavine Perpetual Trophy which is still awarded each year at the Junior League Horse Show in Lexington to the best junior three-gaited horse. The Lexington Junior League Horse Show is the first step toward attaining the Triple Crown of the American Saddlebred Competition. The second competition is the World Championship at the Kentucky State Fair in Louisville, followed by the American Royal Horse Show in Kansas City.

Champion Beau Peavine, with Cape Grant up. From the collection of the American Saddle Horse Museum, Lexington, Kentucky.

A sketch of Chief of Spindletop as done by George Ford Morris in 1934 at the American Royal. Courtesy of Mr. Lynn Weatherman, editor of the *American Saddlebred Magazine*.

Chief of Spindletop, with "Capable" Cape Grant up. From the collection of the American Saddle Horse Museum, Lexington, Kentucky.

Last picture of Roxie Highland just before she was retired from show ring, with Manager Cape Grant, up.

Champion Roxie Highland as painted by George Ford Morris; from the collection of the American Saddle Horse Museum, Lexington, Kentucky.

Senator Crawford at his retirement in Denver, Colorado as it was broadcast across the U.S. Standing in front of Senator, holding one of his trophies, are Mildred and Edward Manion. The announcer is W. Jefferson Harris of Oklahoma. To the left of Harris are Mr. and Mrs. Doc Flanery. Courtesy of Mr. Lynn Weatherman, editor of the *American Saddlebred Magazine*.

Roxie Highland and her chestnut daughter, Roxie Highland of Spindletop, out of Beau Peavine, taken just before Roxie's death in 1939; courtesy of the American Saddle Horse Museum, Lexington, Kentucky.

Roxie Highland's grave marker which can be seen at the Kentucky Horse Park in Lexington, Kentucky. It was moved from in back of Spindletop Hall, although Roxie's actual grave remains on Spindletop Farms.

Legendary Spindletop horse Lady Virginia and her foal Lady Augusta, as taken in 1938. From the collection of the American Saddle Horse Museum, Lexington, Kentucky.

Miss Dixie Rebel. From the collection of the American Saddle Horse Museum, Lexington, Kentucky.

The legendary roadhorse, Senator Crawford, at his retirement ceremony in Denver, Colorado in 1939, with Doc Flannery and Al Pendleton. Courtesy of Mr. Lynn Weatherman, editor of the *American Saddlebred Magazine*.

Original Entrance Gates to Spindletop Hall, courtesy of Kathryn Manion Haider. Photo by Lafayette Studio.

Spindletop Hall, Lexington, Kentucky, courtesy of *Saddle and Bridle Magazine*, St. Louis, Missouri.

Pansy Yount's cherished Coach and Four before Spindletop Hall. From the collection of the American Saddle Horse Museum, Lexington, Kentucky.

Original Foyer of Spindletop Hall with its Winding Staircase Leading Up to the Manion Suite at the Top of the Stairs, courtesy of Kathryn Manion Haider. Photo by Lafayette Studio.

Original William and Mary Music Room at Spindletop Hall, courtesy of Kathryn Manion Haider. Photo by Lafayette Studio.

Original Elizabethan Living Room at Spindletop Hall, also called the "Oak Room," courtesy of Kathryn Manion Haider. Photo by Lafayette Studio.

Original Gothic Library at Spindletop Hall, courtesy of Kathryn Manion Haider. Photo by Lafayette Studio.

Original Georgian Dining Room at Spindletop Hall, courtesy of Kathryn Manion Haider. Photo by Lafayette Studio.

Original Angelica Kaufmann Sitting Room of Mildred's Suite Upstairs in Spindletop Hall, now called the Manion Suite, courtesy of her daughter, Kathryn Manion Haider. Photo by Lafayette Studio.

Original Way Mildred's Bedroom Looked at Spindletop Hall, courtesy of her daughter, Kathryn Manion Haider. Photo by Lafayette Studio.

Original Rose Gardens at the Back of Spindletop Hall, courtesy of Kathryn Manion Haider. Photo by Lafayette Studio.

Original back of Spindletop Hall showing the South Terrace, courtesy of Kathryn Manion Haider. Photo by Lafayette Studio.

Two of Pansy Yount's best Kentucky friends: former Kentucky Governor A.B. "Happy" Chandler on the left and Mr. Fred Wachs, Sr., on the right. Courtesy of Mr. and Mrs. Fred Wachs, Jr.

Mildred Yount in her wedding gown at her wedding reception at Spindletop Hall in 1938, after being married at St. Paul's Catholic Church in Lexington, Kentucky. From the collection of the American Saddle Horse Museum, Lexington, Kentucky.

A painting of Miles Frank Yount by Mr. George Stanhope Wiedemann, Lexington, Kentucky.

An older Pansy Yount. Courtesy of Mr. and Mrs. Fred Wachs, Jr.

(Opposite page:) Russian artist Michael Perovski's painting, "Slave," was one of Pansy Yount's favorite works of art that she displayed prominently in Spindletop Hall. When she was leaving Spindletop Hall for the last time and selling it to the University of Kentucky, she gave Perovski's painting to her long-time Kentucky friend Fred Wachs, Sr.; the Wachs' did not want it, so when U.K. took over Spindletop Hall, Fred Wachs, Sr., gave it back to Spindletop. They apparently did not know exactly what to do with it, either; thus, the painting "almost without a home" now resides in the men's restroom on the lower level of Spindletop Hall. As one of her favorite art works, it serves as an interesting backdrop for interpretation when held up against Pansy Yount's life and experiences. Was it just an art work to her; or what did she see expressed in Perovski's masterpiece?

Michael Perovski, "Slave," Moscow.

Actress/Author Linda Light as Pansy Yount of Spindletop. Photo by Danny Sylvestri, Impact Photography, Lexington, Kentucky.

The brute force and power of an **oil gusher at Spindletop Hill May 9, 1927**, breaks a wooden derrick into tiny pieces.

Aerial photography of the original layout of Spindletop Farms and Spindletop Hall in Lexington, Kentucky. Photo courtesy of Mrs. Ed Fitzpatrick.

An original photo of the foyer of Spindletop Hall in Lexington, Kentucky, when Pansy Yount occupied Spindletop Hall plainly shows that Pansy Yount had her picture on the *first floor* above the fireplace in the foyer as one entered Spindletop Hall. Her picture was removed and replaced by a portrait of Miles Frank Yount, who never saw Spindletop Hall, because, according to Dr. Frank Peterson, someone thought she didn't "look representative enough" to be in Spindletop Hall.

This painting of Pansy Yount has been done and relegated to the second floor of her beloved Spindletop Hall in Lexington, Kentucky.

Pansy Yount had each of Snow White's Seven Dwarfs a special "life-sized" fantasy dwarf house built at Spindletop Farms in Lexington. Featured are builders Junus "Buck" Thompson left and, by the ladder, Ed Fitzpatrick, to whom Pansy Yount eventually gave the diamond ring originally intended for Cape Grant. Photo courtesy of Dan Thompson.

Another one of the dwarf houses for Pansy Yount's collection of figures from Snow White and the Seven Dwarfs that she had built on Spindletop Farm in Lexington, Kentucky.

Pansy Yount brought the first imported Percheron horses to Kentucky. Here Junus Thompson displays one of Spindletop Farms' Percherons. Photo courtesy of Dan Thompson.

Cape Grant with one of Spindletop Farms' prized registered Angus bulls. Photo courtesy of Mrs. Ed Fitzpatrick.

Pansy Yount had a kennel at Spindletop Farms in Lexington, Kentucky, where she raised all kinds of dogs. Pictured above are her prized Llewellyn Setters and Pointers with which she competed with Cape Grant in field trials.

A 1940 photo from *National Geographic Magazine* showing a young Thoroughbred at Spindletop Farms in Lexington, Kentucky, with Junus "Buck" Thompson holding the mother's lead and Lexington veterinarian, Dr. D.L. Proctor, administering castor oil. Photo reprint courtesy of *National Geographic*.

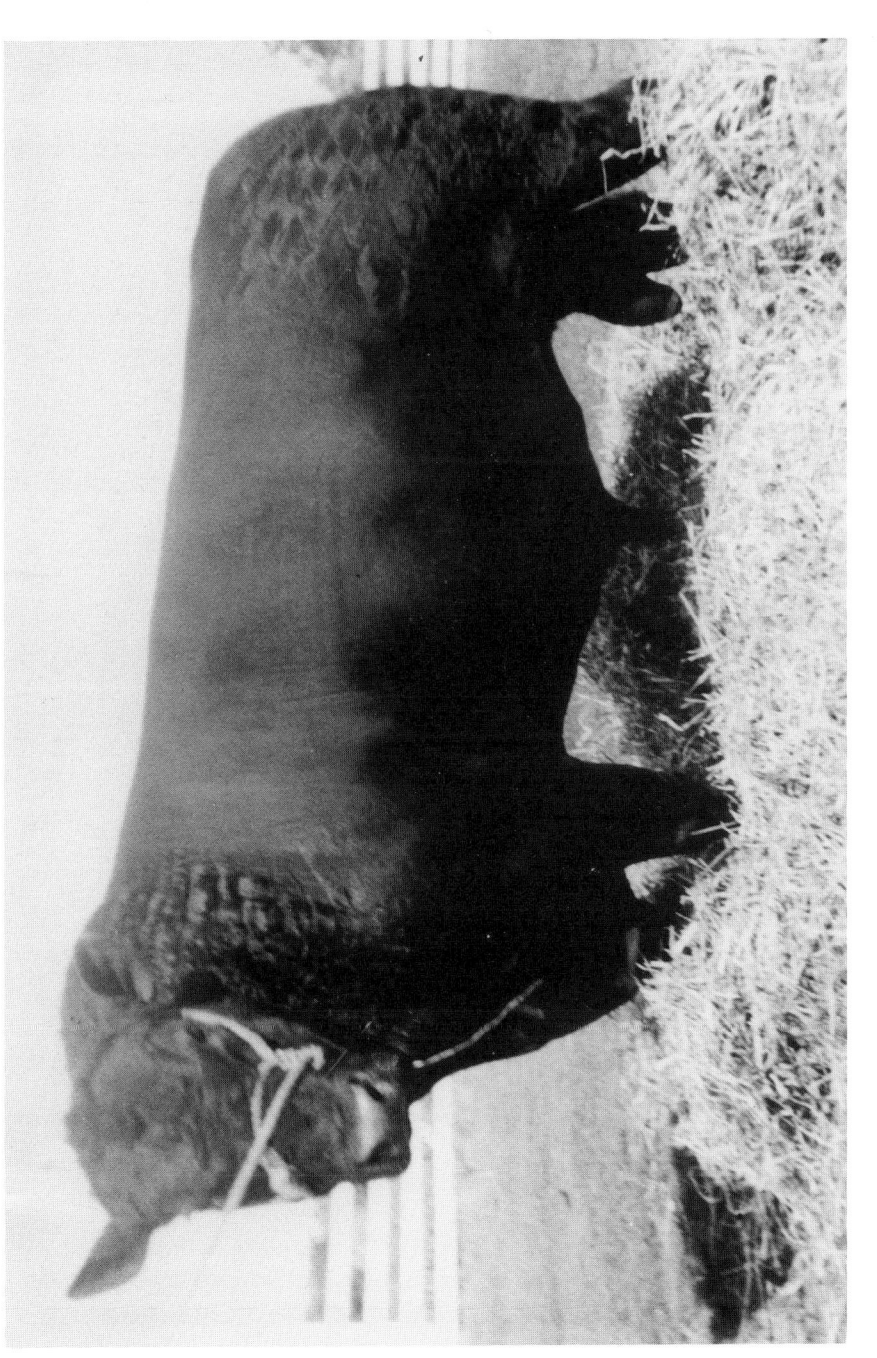

"A-whole-lot-of-bull." Mrs. Yount's prized angus bull at Spindletop Farms in Lexington, Kentucky. Photo courtesy of Dan Thompson.

An early home of Frank and Pansy Yount in Sour Lake, Texas, known as "**Sunnyside**," Courtesy of Bruce Yount.

The beautiful Mansion foyer and staircase at **124 E. Caldwood** at the time Pansy Yount lived there. Photo by Rolfe Christopher, The Christopher Studio, Beaumont, Texas.

The Mansion at **124 E. Caldwood in Beaumont, Texas**, where Pansy Yount died on October 14, 1962. To locals, this prestigious historical location in Beaumont is known as the **Steinhagen Mansion**. 1958 Photo by Rolfe Christopher, The Christopher Photography Studio, Beaumont, Texas.

Rich woods and unique lamps inside the entrance hallway of the Mansion at **124 E. Caldwood** at the time Pansy Yount lived there. Photo by Rolfe Christopher, The Christopher Photography Studio, Beaumont, Texas.

The beautiful Mansion foyer and staircase at **124 E. Caldwood** at the time Pansy Yount lived there. Photo by Rolfe Christopher Of The Christopher Studio, Beaumont, Texas.

The library in the **124 E. Caldwood** Mansion. Pansy Yount loved to read, had over 6000 books in her personal library, and received a commendation from the Smithsonian Institute for her collection of Sherlock Holmes books. Photo by Rolfe Christopher, The Christopher Studio, Beaumont, Texas.

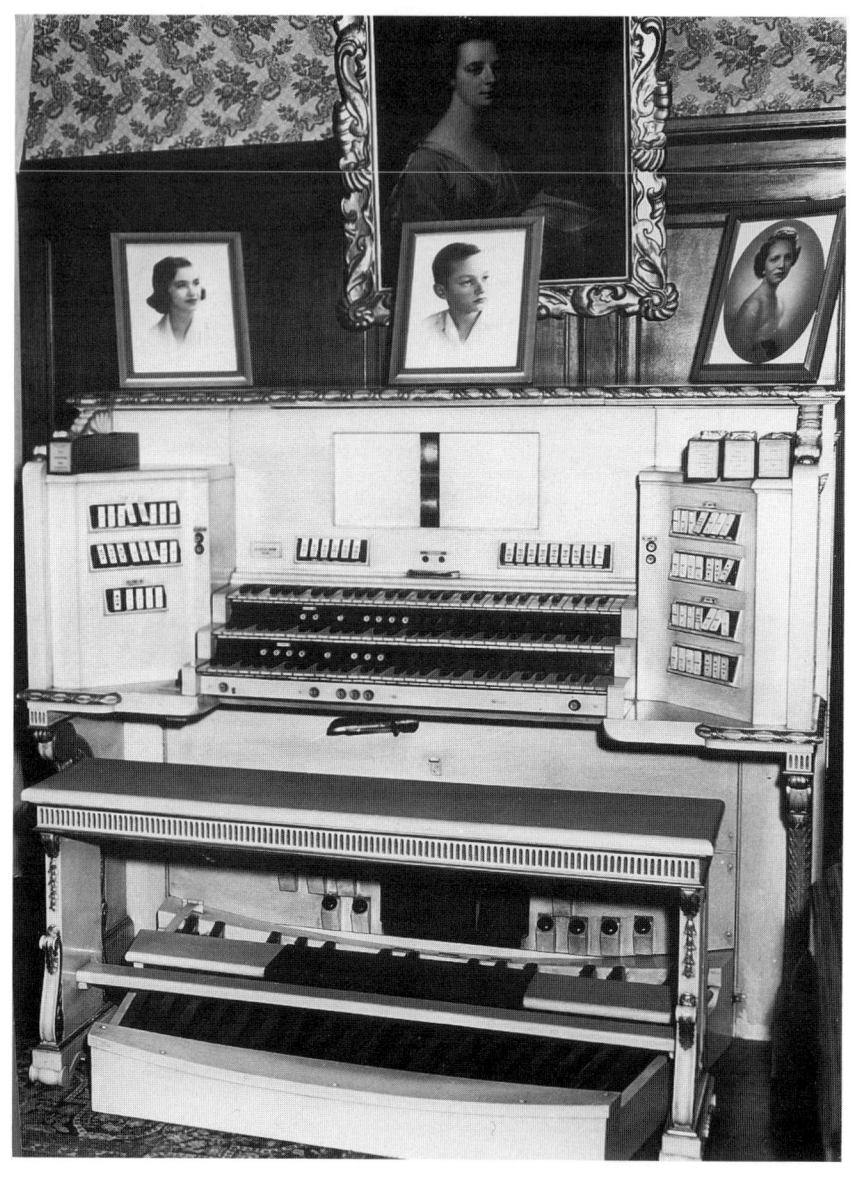

Pansy Yount's organ in the Mansion at **124 E. Caldwood in Beaumont, Texas**. The three pictures on top of the organ are pictures of Pansy's grandchildren, from left to right: Mildred Frank Manion, Edward Manion, and Kathryn Manion Haider.

Pansy Yount's Rolls Royce and cars in her garage at the Mansion on **124 E. Caldwood.** Photo by Rolfe Christopher, The Christopher Studio, Beaumont, Texas.

The servants' quarters at Pansy Yount's Mansion at **124 E. Caldwood**, now a guest house. Photo by Rolfe Christopher, The Christopher Photography Studio, Beaumont, Texas.

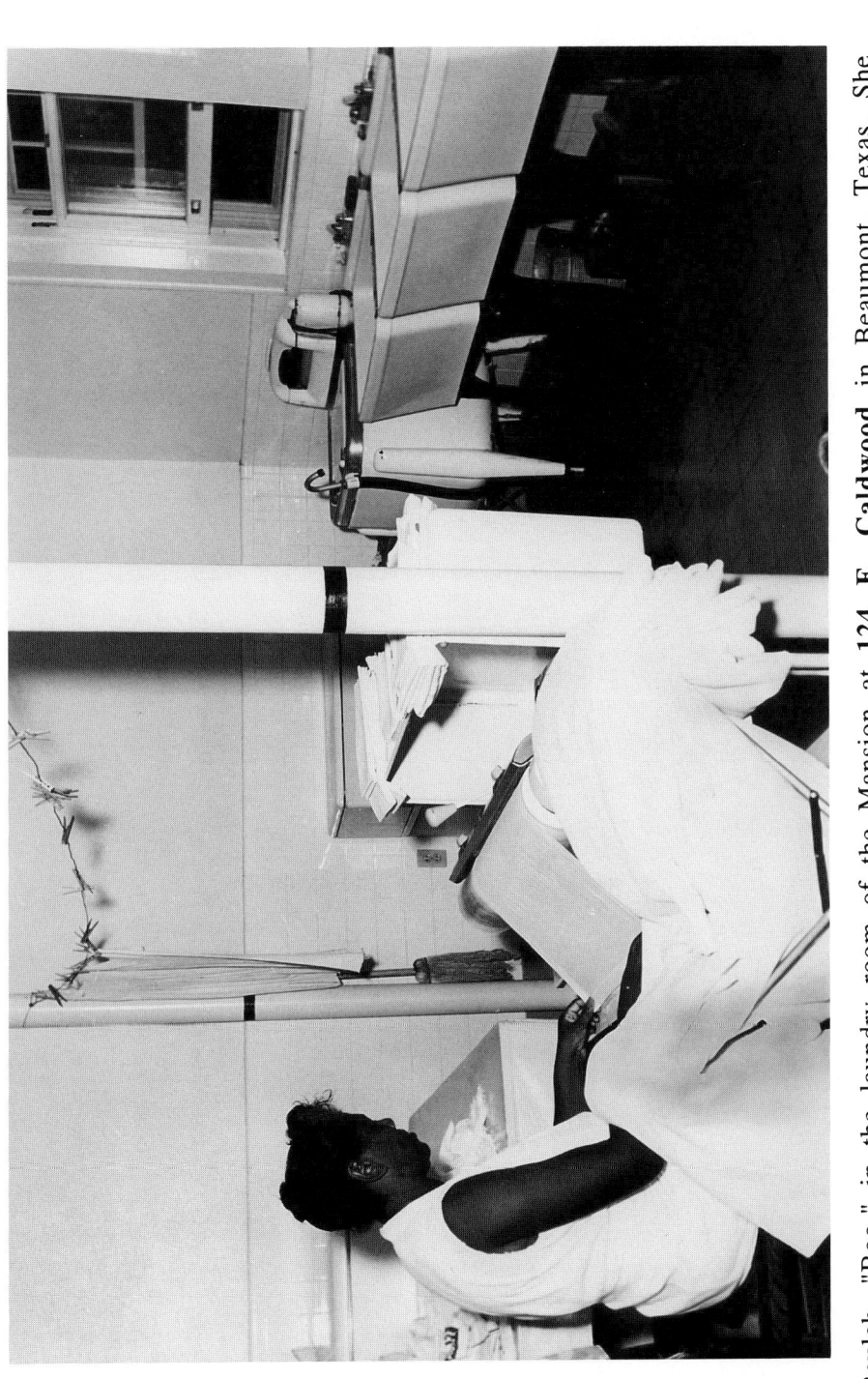

Beulah, "Boo," in the laundry room of the Mansion at **124 E. Caldwood** in Beaumont, Texas. She was a life-long help and friend to Pansy Yount. As Whitey Kahn, past owner of "Le Cheval, Ltd." In Lexington and past groom to Mrs. Yount at Spindletop Farms emphasizes: "the help liked Mrs. Yount. *The help, the farm hands, the grooms, the people who worked for her genuinely liked her.* And that speaks volumes."

The house at prestigious 650 Thomas Road in Beaumont, Texas, where Pansy's daughter, Mildred, and her husband, Ed Manion, moved after their wedding in Lexington, Kentucky. Photo by Gary Christopher, The Christopher Photography Studio, Beaumont, Texas.

Today the Mildred Building, built by Frank Yount for his beloved daughter Mildred, located at **460 Mariposa** in Beaumont, Texas. Today the Mildred Building houses apartments and is still a prestigious address.

Pansy Yount was a devout Catholic and grew up going to **St. Anthony's Catholic Cathedral** (pictured above) located at **700 Jefferson Avenue in Beaumont, Texas.** Photo by Gary Christopher.

The pieta outside of beautiful St.. **Anne's Catholic Church** located at **2715 Calder Avenue in Beaumont, Texas**, where requiem mass was said for Pansy Yount on her death, October 14, 1962, and where she was a faithful member.

Pansy Merritt Yount bought her mother a gravestone and her family a monument at **Magnolia Cemetery, 2290 Pine Street, Beaumont, Texas** with the first money from the Spindletop oil gusher.

The grave of Pansy Yount at the entrance of the new section of **Magnolia Cemetery**, across the street from the original Magnolia Cemetery at **2290 Pine Street, Beaumont, Texas**. The graves of Miles Frank Yount, Mildred Yount Manion, and Edward Manion are located there beside Mrs. Yount's grave.

After the second Spindletop oil gusher--that of Miles Frank and Pansy Yount--came in on November 13, 1925, oil wells covered Spindletop Hill like ants as seen in this 1926 photo. Out of Spindletop oil came such oil giants as **Mobil, Texaco, Gulf, and Sun Oil.**

The Yount-Lee Oil Company's repair gang at Spindletop Hill with C. Hopper, foreman.

Spindletop Hill in 1926 when an oil fire broke out.

Another shot of Spindletop Hill, May 5, 1927.

Texas oil wells today sport the nickname "**Texas grasshoppers.**" Photo courtesy of Charlie Schwab.

A Dollar Bought a Lot in 1930

Pansy Yount was propelled from virtually nothing to the status of a multi-multi-millionaire in the 1930's after Spindletop came in in 1925. In terms of purchasing power for those days and times that would be unfathomable wealth and almost unimaginable, even by today's standards. Today's equivalent would probably be a multi-multi billionaire. To just give an idea to the reader of the buying power of the Spindletop fortune in Spindletop times, consider the following.

Here's what life was like in 1930 according to an adult resident of the time. Just imagine:

> Five gallons of gasoline was 85 cents;
> One gallon of kerosene--18 cents;
> One quart of oil--15 cents;
> A haircut--25 cents;
> A roll of toilet paper--25 cents;
> Telephone bill--$2.25 a month;
> Three pounds of rice--18 cents;
> A gallon of milk--12 cents;
> One dozen eggs--22 cents;
> A bakery-type apple pie--10 cents;
> Two loaves of bread--10 cents;

PASSIONS AND PREJUDICE

Two pounds of butter--25 cents;
Three pounds of brown sugar--21 cent.

According to this same 1930 adult, "I used kerosene lamps, and the stove was kerosene run. We had no electricity. My average income from 1930 through 1933 was $3.00 a week. Our first child was born at home."

Now imagine the Spindletop oil fortune of millions and their purchasing power against those figures.

APPENDIX II

SPINDLETOP HALL AND SPINDLETOP FARMS AT THEIR HEIGHT

SPINDLETOP HALL AT ITS HEIGHT

The following appeared as a room-by-room, article-by-article description of the lavishness of Spindletop Hall, its furnishings, and Spindletop Farms at their heights, just after it was built in Lexington, Kentucky in 1937, by Pansy Yount. The description was published along with pictures first in 1938 in the *Lexington-Herald* newspaper and then in the book *The Enchanted Bluegrass* by Elizabeth Simpson, also in 1938:

LEXINGTON-HERALD, SEPTEMBER 18, 1938

WITH WORLD TO CHOOSE IN, MRS. YOUNT PICKED LEXINGTON

SPINDLETOP HALL PALATIAL HOME

Planned for Comfort, House is Beautiful as Fairy Princess'

...Spindletop remains the most exciting chapter in the history of the southwest.

As God made the west for the years of achievement, so He made the Bluegrass country for the days of rich fulfillment. With all the world

PASSIONS AND PREJUDICE

from which to choose, Mrs. Miles Frank Yount came to Lexington in 1934 and selected as the site for her palatial manor house and the home of some of the world's most celebrated show horses, the eight-hundred-acre tract of W.R. Coe's elaborate thoroughbred establishment, Shoshone Farm, at the intersection of the Newtown and Iron Works pikes.

N.L. Ross, Colorado Springs, was engaged as the contractor, and E.T. Hutchings, Louisville architect, was asked to draw plans for the house, following the clearly defined ideas of its owner. For thirty years, she had planned to build a home, and every detail had been carefully thought out before she consulted an architect. She wanted a home where comfort would out-weigh every other consideration, yet it must be beautiful as any palace of a fairy princess, for it was to be the background for her only child, Mildred Frank Yount, namesake of her husband. It should have mantels and tapestries and rugs from the far corners of the earth, and it should have beds and chairs and sofas of deepest luxury.

Wide wrought-iron gates swing back to give entrance to the magnificent estate down an avenue of maples. Gnomes peer out from behind trees, and "the little people" rest among the perennial borders that line the lawn enclosure.

SPINDLETOP HALL AT ITS HEIGHT

Six large fluted pillars support the portico of the north entrance to the Georgian, red brick mansion, and six pilasters are against the dressed stone wall that lines the face of the verandah. The copper roof has turned a lovely green, and the deep cornice is decorated with Adam swags below the rows of dentils. Wrought iron balconies break the severity of the facade at upper windows, and large arched windows flank the double entrance doors.

The interior is gloriously beautiful. Poised and serene, its splendor of dimensions, exquisite frescoing, priceless rugs, and tapestries, and sumptuous carved paneling take nothing from its atmosphere of ease and enjoyment. It is primarily a home, lived in and loved.

Halbert White, Kansas City decorator, working with Mrs. Yount, has created a masterpiece that has no parallel in Kentucky.

Its fine architectural details are evident from first glance into the great entrance hall, with its graceful, flowing line of stairs rising at each side of the wide doorway. The windows are hung with heaviest Italian silk velvet in a shade of rosy red and the steps are carpeted in red chenille to match the imported carpet of Georgian design, a motif that is reflected in the molded ceiling. Chairs and benches are in the Chippendale manner, as is the table that has a top of Egyptian marble. The stair is

PASSIONS AND PREJUDICE

white, three designs alternating in the spindles, and Corinthian columns supporting the balcony, take on a shell pink glow. A handsome lantern swings from the ceiling.

Near the door of the French powder room hangs a Renaissance tapestry, "The Hunt," woven about 1600, with a wide and intricate border.

The powder room on the left of the hall might well have been a little salon of Marie Antoinette, with its walls of palest flesh pink, the designs of the carving emphasized in gold leaf. The panels are of green silk brocade. A huge French mirror fills the shallow alcove at the end of the room above the dressing table. Gray painted French chairs and sofas are covered with coral velvet and the draperies are of Louis XV hand-woven silk brocade of biscuit color and polychrome, made for the room in Lyons, France. Paintings of great French courtesans adorn the walls, and light falls from a crystal prism chandelier in the center of the room. The floor is laid in parquetry.

Double doors at each side of the hall fireplace open into the Elizabethan living room and Georgian dining room, and a door on the right gives access to the William and Mary music room, the walls wainscoted to the ceiling in walnut. A tapestry hangs above the Eighteenth Century mantel of white marble with Alps green frieze and

SPINDLETOP HALL AT ITS HEIGHT

pilaster backing designed by Kent in 1735, and taken from a house designed by him on Marlborough street, in London, England.

The draperies are a gorgeous William and Mary design woven in heavy silk damask of rich apricot. The valances are all solid wood and all of the surfaces, including the mouldings, have the damask pasted on in exact copies of authentic William and Mary work now preserved in one of the great English museums. A Royal Sarouk rug covers the floor, and the crystal chandelier is a thing of unusual beauty. A little rosewood melodeon, one of the finest examples of these instruments in existence, stands near the door into the living room, and a Kimball organ at one end of the room may be turned on in twelve other places in the house.

A harp of decorated satinwood and gold, designed near the grand piano, and a case built into the fireplace end of the room holds the collection of rare violins that belonged to Mr. Yount. In his fervid pursuit of knowledge in diversified fields, Mr. Yount became interested in violins when his daughter first showed signs of unusual talent. He read everything available concerning artists and instruments and soon became a most discriminating collector, conversant with the many fine points by which the great masters are identified.

PASSIONS AND PREJUDICE

As the foundation of his collection, he acquired two Stradivari. One is the "Reynier," 1681, and the other the "Piatti," 1717. Both are remarkably well preserved with their original varnish, rich in color. Other famous violins in the collection include examples of the work of Dominicus Montagnan, Andreas Guarnerius and his son, Joseph, Joannes Baptisa (!) Guadagnini, and Jacobus Stainer. Fine bows include several by François Tourte, Pecatte, and Jean Baptiste Vuillaume.

A few steps at the left of the fireplace lead down to the library, paneled in Gothic oak, with a high, hammer-beam ceiling. The English Tudor gray stone mantel was removed from Trentham Hall, Staffordshire, one of the seats of the Duke of Sutherland. A wonderful old Brussels tapestry of early Renaissance weaving, formerly the property of Emperor Franz Josef, hangs above it, and the iron fender, its wide top covered with brown leather, forms a fireside seat.

Parchment lamp shades are copies of a Blanche of Castile tapestry in the J.P. Morgan collection and a Charles VII tapestry in the Metropolitan. Oriental rugs of almost unbelievable worth are strewn over the floor, for Mr. Yount was also an avid collector of rugs and rare editions, and became an authority on them. Books on shelves that line the walls are reached from a spiral library ladder.

SPINDLETOP HALL AT ITS HEIGHT

The Elizabethan living room, thirty by sixty feet in dimensions, with walls of Gothic oak carved in parchment-fold panels, has a frieze of three alternating designs of Elizabethan strap work.

The ornamental plastered ceiling is reproduced from an Elizabethan room in an old English country house. The elaborately carved mantel bears the inscription, *"East, West, Home's Best."* A rug nearly 300 years old hangs above it. A green seamless chenille rug of luxurious depth of pile covers the floor to the baseboard.

A stringer table holds a silver punch bowl, a duplicate of the Dixiana perpetual trophy won by Chief of Spindletop in the 1936 Louisville horse show the night he became the five-gaited world champion.

The dining room, with its walls of cool Adam green, is resplendent with five crystal chandeliers, prism sidelights and touches of gold-leaf outlining the panels. The rug, woven for a room in Czechoslovakia, is in softest shades of pearl gray, rose, buff, and blue greens, and the Georgian pattern of scrolls and flowers is repeated in the moulded ceiling. "Spindletop Hall" is woven inconspicuously into the narrow border in front of the fireplace, and the initials of Mrs. Yount and her daughter are faintly discernible in the corner scrolls.

PASSIONS AND PREJUDICE

The fine old mantel, made in 1750, was imported for the New York mansion of Otto Kahn and sold when the place was dismantled. It is fashioned of Cararra marble with twin black and gold marble columns at each side, and a convent Siena frieze inlay. It originally adorned Shapwick Hall, Somerset, England.

Atop a decorated linen cabinet is a Royal Copeland platter. The dining table, with its Egyptian marble top, is one of a string that can be made to seat a large number of guests. A silver closet, glass enclosed, fills the end of the room opposite the fireplace, and doors on either side opening into the butler's pantry are opened by electric "eyes."

The draperies are of superb Royal blue and gold Louis XV Chinoiserie designed by the famous Pillmonte.

Above a long serving table is a Seventeenth Century Royal Gobelin tapestry, twelve feet wide and eleven high, delineating the episode of Carlo and Ubaldo at the Fountain of Laughter in the Romance of Rinaldo and Armeda, from Tasso's "Jerusalem Delivered." The tapestry formerly was in the collection of Clarke-Thornhill Esq., and hung in Rushton Hall, the beautiful Tudor castle in Northamptonshire, England.

SPINDLETOP HALL AT ITS HEIGHT

Both the living room and dining salon open on the south side of the mansion with its broad terrace, lacy wrought iron balustrade and pillared circular portico. The porto cochere, with decorative iron gates, is approached from the library. All outside doors are of bronze with grille fretwork.

The service end of the house includes a white-tiled butler's pantry, its shelves holding Wedgwood, Royal Doulton, and Dresden, hand wrought silver and delicate crystal. A salad pantry adjoins it, and in the large white kitchen, Will Irvine, chef in the service of the family for twenty-five years, rules supreme. From his enthusiasm one gets a perfect picture of the master he served so devotedly, and of the mistress who hold the affection and deep loyalties of those in her employ.

The staff dining room, service stair, flower room, and pantries open on the rear hall.

Mrs. Yount's suite is on the second floor, opening off the balcony opposite the entrance doors of the north terrace, and its windows look out on the sunken rose garden.

Done in the period of Louis XV, its oyster white paneling is on walls of buff, with a deep moulded frieze. The fireplace is of dove gray marble and is one of a pair taken from an ancient

PASSIONS AND PREJUDICE

French chateau. Above the mantel is Nicholas Largilliere's portrait, "Lady of the Court." Draperies are of gold and silver satin damask, and the chandelier and side lights are French gilt mesh interspersed with pastel enamel flowers. French portraits, divans and chairs upholstered in rose velvet, a French cherry commode, and a pale ivory baby grand piano, artistically decorated, give the room an air of delicate charm. The floor is carpeted in French blue. Benell portraits of Mrs. Yount's parents, Hosea Holly Merritt and Sarah Frances Sherman Merrit, hang at the end of the room.

The adjoining bedroom in the mode of Louis XVI, has twin French beds draped with blue and putty damask of authentic Louis XVI design, and oval miniatures of Mildred Yount as a little girl hang just above the low headboards. Rose pointe lace covers the lamp shades, and a French mirror hangs over the dressing table. Near the fireplace is a George Ford Morris sketch in oil of "Tiny," once a favorite Pomeranian member of the household.

On the shelves of a French cabinet are treasures suitable for museums. There is a lorgnette of blue enamel and diamonds that Napoleon gave to his beloved Josephine; a watch with a fine miniature in enamels that once belonged to Alexander II of Russia that was given to Mr. Yount by a Russian prince; a set of Spanish earrings, comb and necklace of cameos set in gold

SPINDLETOP HALL AT ITS HEIGHT

filigree brought out of Madrid during the revolution; ...and a star sapphire ring of such size and color as to represent a king's ransom.

The dressing room adjoining the bedroom, its walls filled with hanging closets, was decorated by Dennis Prectyl, Cincinnati artist, and forms the connecting link with the bathroom that is tiled in green with mirrored wall above the dressing table and the tub in the center of the room.

Adjoining Mrs. Yount's suite is that of her daughter, Mildred....It is said to be the only complete Angelica Kaufmann apartment in America. Its satinwood and cane furniture was faithfully copied from pieces of late Eighteenth Century period in England and decorated in the manner of the great Kaufmann.

A baby Steinway grand piano stands in strong contrast to the smallest of fine and ancient melodeons. A portrait of "Mrs. Perry," by George Harlow, hangs on one wall, and nearby stands a silver trophy that was presented to Mildred Yount by the city of Beaumont the night that Roxie Highland, her national champion five-gaited mare, was retired at Madison Square Garden in 1935.

Holding a place of prominence in the room is a pleasing portrait photograph of Mr. Yount in academic robes as a regent of the University of Texas.

PASSIONS AND PREJUDICE

As heiress to the largest fortune in the southwest, the idolized only child of the house, Mildred Yount Mansion, not yet in her twenties, is sweetly thoughtful and considerate of others, unspoiled, and touchingly appreciative of even the smallest tokens of friendship. Utterly unaffected, eagerly devoted to out-door life, educated at one of the smart and exclusive boarding schools of the west, she is charmingly fitted to take her place at the head of the beautiful Yount establishment, El Ocaso, in Beaumont someday.

The dressing room of the suite, with closets painted in pastel colors, has diaphanous draperies, and slender, high-backed chairs are covered in oyster white damask. The blue-tiled bath, with tub in the center and mirrored side wall, has glass-enclosed showers, and the light fixtures have pear-shaped prisms.

Four guest suites are on the second floor. One is a replica of Mr. Yount's room at El Ocaso, with Spanish furniture, Oriental rug, red damask hangings, and Paul Doering water colors.

Another is paneled in cream with green carpet, lavender taffeta draperies, dainty dressing table and chaise longue.

One, in shades of pink and blue, has walls adorned with F. Kaufmann seascapes, and the adjoining bath is tiled in pink and cream.

SPINDLETOP HALL AT ITS HEIGHT

A single Italian bedroom with bath, the apartment of Mrs. Yount's secretary-companion, is in tones of wisteria and green. And two early Kentucky bedrooms and bath are on the third floor.

Stairs descend from the great entrance hall to the saddle lounge in the basement done in worm-eaten chestnut paneling, and blue and red leather, the colors of Spindletop Stables. The massive chimney is built of greenstone and redstone quarried on the Yount property in Colorado. It has faint coloring of green and pink and is a touch of the West added as a tender bit of sentiment. A large buffalo head hangs above the opening of the fireplace. Shades on the lamps are copied from century-old Maryland hunting scenes.

On the walls are stag heads and George Ford Morris paintings of Beau Peavine and Chief of Spindletop, and crayon sketches by the same artist. Photographs of Senator Crawford, champion road horse, and Calumet Armistice, both of the blood of Peter the Great, are hung alongside the trunkload of ribbons won at the Chicago Jubilee Show.

Carriage lanterns are used in the lighting fixtures, and a table is ingeniously fashioned of cart wheels and singletrees.

PASSIONS AND PREJUDICE

The men's room, opening on the left, has walls decorated with wisteria blossoms.

The Kentucky tap room is a fascinating replica of an old wayside tavern. It, too, has a fireplace of Colorado stone, and paneled walls. The bar has a brass rail, and bar-room slogans are framed around the mirrors. The high-seated chairs are the western bar-room type with spotted ponyskin and cowhide cushions. A large front window with many small panes is flanked by doors opening on a court which comprises the ballroom, the verandas of the tavern enclosing it. The dance floor is constructed with a slight give to prevent the dancer's tiring. The powder room has silver walls adorned with cheery blossoms and hung with Chinese dolls in costumes of the various dynasties.

The dog room, arranged with bath facilities and decorated in typical canine manner, is occupied by the Pomeranian house dogs and a prized Griffon, one of only two or three in America, as they are seldom able to stand the ocean voyage.

At the right of the garden is the large swimming pool with chute-the-chute and white terrace furniture, and the brick bathhouse has a colonnade center designed for al fresco tea. A tennis court, built of composition and painted

SPINDLETOP HALL AT ITS HEIGHT

green, lies between the bathhouse and the stable for pleasure horses, with the kennels beyond.

Architects fairly tore their hair at the proximity of the stable and kennels, but to no avail. Mildred Yount wanted her riding horses near enough the house to go out and saddle them herself, and her wishes were law from the time she was able to prattle baby-talk. She and her father were closely companionable. They rode together almost daily, and no business was ever important enough for him to break an engagement with his daughter. So her mother has carried on in the same way, and Della, Mildred's first pony, along with My Mary, her choice mount, is stabled in plain view of the pillared portico. Mrs. Yount's riding horse, Paris Grand, has a stall there, too, as have several other riding horses provided for guests. A Woodward park coach recently has been added to the collection of ancient vehicles.

In the kennels are Scotties and Sealyhams, Great Danes, Llewellyn setters, Dalmatians, pointers and foxhounds, each breed with its own run. Electrified doors of kennels and dairy sizzle every fly that aspires to enter.

Spindletop Farms, however, is more than an elegant and costly whimsy with Mrs. Yount. It is conducted as a sort of experimental farm where various breeds of stock and fowls are being perfected under the best possible conditions.

PASSIONS AND PREJUDICE

Hampshire hogs, a registered Aberdeen-Angus herd, a flock of Shropshire sheep, and mild goats are carefully bred and culled. The Jersey dairy herd is headed by Kahokas Dream Not, imported from the Isle of Jersey and winner of several championships. Adams Model, international prize winner among Percheron horses and the first imported Percheron brought to Kentucky, is in stud at Spindletop.

White Leghorn chickens, turkeys, peacocks, ducks, geese, pigeons, and bantam chickens are in the various enclosures.

The service court, with garages and servants' apartments, is at the extreme left end of the mansion, and the greenhouse, within a stone's throw, is filled with roses and violets, camellias and gardenias and chrysanthemums.

A grove of trees in the midst of which a barbecue pit has been built, embowers lovely Lake Mildred which covers an area of twelve acres. Black swans mingle among the snowy white ones that glide on its glassy surface, and small boats lie at anchor near the banks. Another lake five acres in size and known as Lake Roxie is on another part of the farm, an island in its center offering a sanctuary for birds.

Mrs. Yount is passionately devoted to horses, but the international success and fame of the stable is due also to the remarkable ability of W. Cape Grant, manager and trainer. Few men can handle show horses so unfalteringly, and few are born with the showmanship that is his. He fairly vitalizes the tan-bark oval the minute he enters and flashes his famous smile. It was true the night he rode Chief for the world's championship state at the Louisville show in 1936, defeating Dixiana's Night Flower, the former champion, with Charles Dunn up. The victory was seen as a triumph for the sole trainer and rider as well as for the great gelding himself.

The Chief is still one of the best geldings out, a beautiful chestnut eight-year-old that has won in all leading shows. But Beau Peavine, by Jean Val Jean out of Fair Acre by Vanity Fair is the pride of Spindletop, winning under saddle, in fine harness, model, and breeding classes. He is one of the premier stallions of American, as beautiful a specimen as the breed has ever produced, and he has the potent quality of transmitting to his get his own fine points of good head and ears, intelligence, beauty, and way of going....

But Spindletop has withdrawn from the shows. With an international reputation made within the incredible period of five years, the stable will continue to breed its great lines, showing only mares and yearlings with possibly an occasional horse at Louisville, but adhering in the

PASSIONS AND PREJUDICE

main to its announced policy of not competing against owners who buy Spindletop offerings.

Spindletop Farms has eleven barns containing 108 stalls. There are eighty or more paddocks and twenty-three miles of white five-panel fences threading the emerald pastures and woodlands of oak and ash.

Mrs. Yount enjoys the stables and paddocks, the poultry yard and kennels, but she is especially interested in the dairy herd, and on all her farm tours she is invariably accompanied by Lucky, a toy Pomeranian, who takes his business of body-guard with absurd seriousness.

She is instinctively kind and impulsively generous, entirely devoid of artifice and with a fine appreciation of beauty in every form. With loyal courage she stood shoulder to shoulder with a man whose memory is loved the length and breadth of Texas.

They were friends with a deep understanding of each other--companions--"dreamers on horseback"--.

APPENDIX III

Reality Bytes from Spindletop Times: Select Oral Histories

SELECT ORAL HISTORIES

Interview with Mr. and Mrs. Fred Wachs, Jr., (Fred and Dorothy).

Tuesday, July 26, 1994
The Idle Hour Country Club
Lexington, Kentucky

Interviewer: Linda Light

Subject: Pansy Yount of Spindletop, Cape Grant, Spindletop Farms, Spindletop Hall

(Mr. Fred Wachs, Sr., owner of the *Lexington Herald Newspaper* handled in large measure the securing of Spindletop Hall for The University of Kentucky from Pansy Yount. It could be well argued that he was her best and closest friend during her time in Kentucky, and certainly her confidant in which she put full trust.

Mr. Wachs' son, Fred Wachs, Jr., and his wife Dorothy knew Pansy personally and through Mr. Wachs' father.)

Dorothy Wachs: We talked about the bracelet Pansy gave to me. Well, there it is. (Showing bracelet) It's like a charm bracelet. It's very simple and it jiggles, and it makes too much sound that I hardly ever wear it anymore. But she collected these. They are all medals, Catholics medals. She collected them from all over Europe and she sent it to me when, uh.... These were the left-overs that she had made after she had her own bracelet. (Laughs.) I think that's what she told me. Anyway....But they are all gold medals, and she sent it to me when--(not completely intelligible here)--Ben's mother was born, Kathy. I thought it was very sweet and thoughtful.

Interviewer: She was, of course, Catholic. She even tried to give Spindletop to the Diocese of Covington before giving it to the University of Kentucky.

Fred Wachs: They actually signed the contract.

Dorothy Wachs: Fred's father was the contact on that, too.

Fred Wachs: You know when this came up, I started, I remember I had in the house we had been in before the house we are in now, I had a lot of letters and things about the sale of Spindletop Farm. Uh, but the Bishop actually bought, I mean he signed the contract to the farm. And I don't know what happened that he had to back out. There are some letters from Father Joe McKenna here in Lexington.

SELECT ORAL HISTORIES

Interviewer: McKenna was one of her priests.

Dorothy Wachs: He was a dear, dear man.

Fred Wachs: Yeah. Yeah. And he was really upset that the Bishop had backed out of the sale. And then they--I remember he entertained Harold Lloyd, the movie actor. Lloyd, well, he came here. He was president of the national Shriners at that time, and I know that he was out at our farm and they went over to Spindletop, went all through the house, and he was thinking of buying it for the Shriners' Crippled Children's Hospital.

Dorothy Wachs: But I think the reason for the Bishop's not buying Spindletop was complex.

Fred Wachs: I don't know. But I know--but the contract was there, and the Bishop had signed the deed--I mean contract for the sale.

Dorothy Wachs: You know, it was a gift in many ways because, uh,-- but there was some money exchanged, and I imagine it was for tax purposes for her. You know, say, if she sold it for a million dollars and it was worth twenty million-- you know that was a big loss. It would be to me. And, uh, that was sort of the involvement.

Fred Wachs: I was trying to think when we first knew Mrs. Yount. It seems like to me, the

PASSIONS AND PREJUDICE

first time I went out to Spindletop, I went out there with my father to go fishing. You know they had some--some--several lakes out there. One they couldn't get to hold water. Then they had another one--just where they built the building for the Research Center. Of course, the lake then was even bigger than it is there now--before they built that building. We fished in it, I remember.

And I really remember Cape Grant, her farm manager. I remember him before Mrs. Yount, really.

Dorothy Wachs: He wasn't the farm manager; he was the horse trainer. The farm manager was Fitz....

Fred Wachs: That's right, horse trainer. Fitzpatrick was the farm manager.

Interviewer: What kind of guy was he?

Dorothy Wachs: Cape Grant?

Interviewer: Yes, Cape Grant.

Dorothy Wachs: How many horse people do you know? Horse trainers?

Interviewer: Not many.

Dorothy Wachs: (Laughs) Well, I thought he was like a lot of horse trainers, uh--

SELECT ORAL HISTORIES

Fred Wachs: He was kind of a "cowboy"--

Interviewer: Cowboy.

Dorothy Wachs: Cowboy horse trainer.

Fred Wachs: He had a real red face, and, uh--. And he came out to our farm one time when we were sorting calves, and he showed us that he could throw a lasso under the fence and catch a calf by the leg. He really could!

Interviewer: Really?

Fred Wachs: Yeah, he was good. I think he'd been a regular cowboy at one time. Well, he was from Texas--from Texas.

Dorothy Wachs: He always wore a diamond horseshoe ring, or a stick pin or--that type of thing.

Fred Wachs: He liked to bird hunt a lot and kept a lot of high class bird dogs. He had pointers and setters, but I know later--later on when they got a divorce, him and Pansy, Mrs. Yount would try to beat him in these field trials. I mean, she went to England and got the best bird dogs she could find.

Dorothy Wachs: Llewellyn Setters.

PASSIONS AND PREJUDICE

Fred Wachs: Yeah, they were all Llewellyn Setters. And she'd enter them--these dogs in these trials, you know, to try to beat him, I think. And she gave me a pup--a Llewellyn--just before she died. I got the pup--it wasn't six months before she died. Uh, and they were shipped from England. And I had a letter from the Llewellyn game keeper. Of course, old Llewellyn himself was dead, but this was his....uh,--

Interviewer: Was this a surprise?

Fred Wachs: No, no, she just told me she would do it--she was going to do it--because I enjoyed bird hunting and she said, well I'll get you a good dog. So she had it shipped over from England. It really didn't turn out (laughing) to be a very good dog. But I had one of its pups who did. One of the pups turned out to be real good. It was a little hard-headed, but a beautiful dog, black and white Llewellyn--but the pup turned out to be a much better dog.

Dorothy Wachs: We had never been interested in showing dogs and these had been bred to show, you know, so they weren't really working dogs.

Interviewer: What else do you know about Cape Grant?

SELECT ORAL HISTORIES

Fred Wachs: They had an awful divorce. We even had all the depositions taken at the divorce proceedings at one time. My son had all of Dad's letters, and I told him, and we're going to try to find them for you. But I remember, we had all those things. It got to be a rather bitter divorce. Uh, we've got pictures of Cape Grant.

Dorothy Wachs: Paintings. How's that.

Fred Wachs: He talked my father into raising Angus cattle. And I know he arranged to get him some bulls and things like that, and we have pictures of him with some of the bulls.

Dorothy Wachs: And remember when Miss Pansy gave Pap that Jersey cow. The cream was this high on the milk. (Indicates about an inch and a half with her fingers). Oh, we drank it and loved it. What did he call her?--Something like Spindletop's Blossom, or something like that. (Laughs.)

I'll tell you about one night that she came out to Fred's mother's and father's for supper. And we lived with them at that point. And we had, uh, just fried chicken on the back porch, nothing fancy. No air conditioning.--And they came out in a Rolls Royce. I don't know what the chauffeur did-- he never showed up--but Cape and Miss Pansy-- well, we all had supper on the back porch. And she had a bracelet on with big square-cut

285

PASSIONS AND PREJUDICE

diamonds all the way around and then it was framed in many other stones, and she was a wreck. She had gotten it out of the bank vault for some reason--that's really where everything stayed. She didn't much care. She wore a horse pin a lot and long, dangly earrings with sapphires and diamonds.

But she really didn't care that much about jewelry. I understood that Mr. Yount had and had given her--she said he had lots of things made for her.

Fred Wachs: I think she was so worried about the value and wearing it around and everything--

Dorothy Wachs: I think she was worried about somebody hitting her in the head for it.

Fred Wachs: And I think they had to leave early or wanted to leave early and get it back to the bank somewhere to put it back in the vault--

Dorothy Wachs: Or to get it back to the farm, at least.--And here she was, she had two men, the chauffeur and Cape Grant and she was just really upset about the bracelet. She wished she hadn't of gotten it out. *It was elegant*, I mean *elegant*, but as she was leaving she said, "Just don't get too many things--don't collect too many things.

SELECT ORAL HISTORIES

They will wear you out--they will just do you in." I thought that was sad. Wasn't it?

You should have seen Spindletop before she left to go back to Texas. Of course it was full of fabrics and rugs. I don't know what Miss Pansy's taste was, but I rather think it was poor. But that was, you know, judgment. Bad judgment maybe. But she certainly knew where to go to get good advice. And she had really elegant, elegant, elegant things--beautiful--fabrics, carpets, the whole house was lovely. Everything there, every valance over the windows, every piece of furniture had its own linen slip covers--different colors, but piped in different colors--the linen. Most houses don't have that much invested in them. Oh, oh--and runners on the floors--canvas runners.

Fred Wachs: I think every time we went over there, most of the furniture was covered. But she had white runners on the floors to protect the rugs, and then a lot of the furniture had--I call it--dust covers. And she would tell us the price of everything in the house--everything. Every picture. Now this picture cost so much, and this one so much. (Chuckles.) She just seemed happy to have it. Had some marvelous paintings.

Interviewer: Obviously, Pansy and your father were very good friends.

PASSIONS AND PREJUDICE

Dorothy Wachs: They were. They really were. She trusted him. And I must say he went out of his way for her.

Fred Wachs: I'd say he probably started out getting her tickets to things. You know, she'd go to New York and want to see some shows and he had good contacts with an advertising agency there that the newspapers used and he could get her tickets, you know, to any show she wanted to see. I think that's how they kind of met originally.

I can remember him--this was sometime before the war--some Senators came through Kentucky and Mrs. Yount had them out there for lunch. And had my father out there with them. But Harry Truman was one of them.

Dorothy Wachs: I can't remember why they were there.

Fred Wachs: I don't know either. But I remember one of them was Harry Truman. He was a Senator them. I remember my father came home he said he had lunch with a bunch of these drunk Senators. He said they were all drunk but one of them. And he said I don't ever understand how he would have ever gotten to be Senator. And he was talking about Harry Truman. He didn't think--

SELECT ORAL HISTORIES

Dorothy Wachs: He didn't drink anything--at that luncheon. I don't know if he drunk or not. (Laughs.)

Fred Wachs: Cape Grant had us out a number of times---for cookouts--to cook steaks. Of course, he had these steaks and we'd cook out over a fire.

Dorothy Wachs: Sometimes Pansy would be there, sometimes not.

Fred Wachs: I don't think she cared so much for the outings. She'd rather eat at the house somewhere.

Dorothy Wachs: She was never able to keep very good help. They did not suit her, from maybe early on. But as far as house help--cook--had problems with that.

Fred Wachs: Well, I think a lot of them were afraid of Cape, too.

Interviewer: Why?

Fred Wachs: Well, he carried a pistol around. You know, he would--I think he kind of had a hot temper. And, uh, I think they were just afraid of him.

PASSIONS AND PREJUDICE

Dorothy Wachs: I don't think he was that bad. But I think he was really bossy. But, you know, she finally did over her little house, the little farm house, but that was because if she wasn't going to keep Spindletop Hall in proper order she didn't want to use it. And I really don't remember her having anything out there besides the Senators' dinner that time.

Interviewer: You mean, she didn't socialize very much?

Dorothy Wachs: Well, when she first came and first did the house, she had a large party for Mildred. Warren went, Fred's older brother. They were Mildred's age group. For that there were tons of people, but I think most of them went with the attitude of, "What is this old woman doing with all this wealth and elegance." But I think she had her feelings hurt early on.

Fred Wachs: We always heard there were some local ladies went out, and she had an intercom system--and she heard them say something. I've always heard that, I don't know their names.

Interviewer: I understand that she was making a donation at that time.

Dorothy Wachs: To the Junior League--the Junior League of Lexington. Of course, I was busy

SELECT ORAL HISTORIES

with the Junior League and she was always advising them. But she never failed to say, "Now, what you girls need is a big arena inside. This weather is just terrible to show in." We always had rain.

Interviewer: I understand she supported, pushed for the Junior League.

Dorothy Wachs: Yes--but, well, she never volunteered the building. (Laughs.) --That I know of.

Fred Wachs: Oh, I think she was generous to something she was interested in. And she was certainly interested in show horses. The horses won everything. She gave my father two paintings of some of her top show horses, Beau Peavine and Chief of Spindletop. And the only reason she gave them to him was that Cape Grant was up--he was riding them. She didn't want to keep anything around with him on it.--After the divorce happened, you know.

Interviewer: Do you still have these pictures--snapshots or paintings?

Fred Wachs: Yes. These were George Ford Morris paintings. You mentioned Fitzpatrick. One thing I can remember about him. During the war, my father had a tenant on the farm, that kind of a rough fella. He finally decided he was going to

PASSIONS AND PREJUDICE

have to get rid of him. Uh, I don't know, he had had a wreck over on Athens-Boonesboro road and he beat some man up pretty bad. You could hardly get along with him.

Dorothy Wachs: Well, he threatened your father, too, didn't he.

Fred Wachs: Yeah, he threatened him. He was riding a horse on our farm one time--and my father saw him behind a tree. And he kept hiding behind this tree right the way he was riding. So he kind of came around him. Later on he got to worrying about it. He told the Chief of Police about it, and he said, well, get you an ax or something--carry a gun, and if he attacks you, just shoot him and throw the ax down there and say he attached you. Well, at this time, this fella had a child and everything and he didn't want to do that so my father was talking to Cape about it. And Cape says, "Don't worry about it." He says, "I'll tell you what I'll do. I'll send Fitz--Fitzpatrick--out there and let him be your farm manager for about two weeks.

"Well, Fitzpatrick," he told my father, he said, "You know, I got him out of the penitentiary." He said, "He'd killed a man." And he said he had never stolen anything or been dishonest, but this was some man on a farm--I don't know what the problem was, but he had killed this man. But Cape got him out and had agreed to keep him on this

SELECT ORAL HISTORIES

farm. He said, "I'll send him out there" and, he said, "he'll take care of it." He said, "He won't ever bother you again."

Well, my father got to thinking about that. And he thought, Lord, Fitz might come out here and kill that man. And he's got a wife and children, and he didn't want that to happen. (Laughs.) So he finally talked this fella out. But when he left, he tore the door out of the house--out of the tenant house he was in, left the tractor running out in the field, the motor running and the tires turning. Uh, but that was Cape's idea. He thought he would take care of it that way.

Interviewer: Was Pansy Yount accepted here?

Dorothy Wachs: Well, I think it was fifty-fifty. I think she kind of withdrew after awhile.

Fred Wachs: Well, I think she felt that--must have thought that the people kind of looked down on her maybe.

Dorothy Wachs: I think she probably did feel that way. She kept a companion. She had her clothes made in Kansas or some place.

Interviewer: Was this companion's name Ransom?

PASSIONS AND PREJUDICE

Dorothy Wachs: Yes, Ransom. She was just always there.

Fred Wachs: A lot of times Mrs. Yount would come down, and not even stay over at the big house, but stay at the little house on Iron Works-- and later on, Cape Grant's brother lived in that house, I believe. In fact, I saw him just last fall at the races. Jack.

Interviewer: Did you know Cape remarried his first wife just before he died?

Fred Wachs: No. I wasn't aware of that. I know that Cape started drinking awful heavy during this divorce with Pansy. My father still had contact with him. Sometimes, some of the contact was pretty unpleasant with Cape for awhile right up to the time of the divorce. Because, I think, in the divorce proceedings, my father stood up for Miss Pansy. And Cape didn't like that at all. I remember there was a letter that he had written to Cape where he was trying to tell him that after all he and Miss Pansy had been friends a long time and that it was natural that he would take Miss Pansy's side in this thing. Cape, I think, knew himself that he had been, uh--you know, he had gotten drunk and behaved pretty badly with Miss Pansy. And I think before he died, there were several letters where my father and he had pretty well made up.--But I really wonder when he remarried his first wife--.

SELECT ORAL HISTORIES

Interviewer: Two hours before he died.

Dorothy Wachs: I think he was kind of ugly about his sons. One of them wanted to become a doctor and he wouldn't foot the bills or something.

Fred Wachs: I don't ever remember him talking about his sons--his family in any way.

Dorothy Wachs: He sort of took Eddy, Mildred's son, and would ride him around the farm. You know, riding around the farm and so. I think Cape Grant may have had a pretty good relationship with Pansy <u>before</u> they married. (Laughs.)

Interviewer: Who? Pansy?

Dorothy Wachs: Well, yeah. You know, uh, probably kept his cool. (Laughs.) Behaved. --And he was a good horseman. They won everything-- her horses, his managing.

Interviewer: Why do you think they married, and why do you think they divorced?

Dorothy Wachs: I mean, I don't know. I was really surprised at both--the marriage and the divorce.

PASSIONS AND PREJUDICE

Fred Wachs: I think it was kind of convenience in a lot of respects. I don't think Miss Pansy would let him sleep in the house there, you know, unless they were married.

Dorothy Wachs: Well, he did.

Fred Wachs: Well, he did, but it was after they were married. He didn't stay in the house until they were married.

Dorothy Wachs: But he had a bedroom upstairs on the left--

Fred Wachs: Yeah, he did. But that was after they were married, Honey.

Interviewer: Did you have a different impression, Dorothy?

Dorothy Wachs: Yeah, I did. I thought he was in there before they married.

Fred Wachs: I don't believe so. Not Miss Pansy. I think she wanted somebody in there really kind of for protection. All her possessions and everything.

Interviewer: And why do you think they divorced later on. He filed first, and then she filed a countersuit.

SELECT ORAL HISTORIES

Dorothy Wachs: I think that's when he started his drinking and got sort of--

Interviewer: He seemed to have a reputation for drinking.

Fred Wachs: Yeah. I think when he started drinking he, you know, got pretty mean--

Interviewer: Do you think Cape Grant was alcoholic? Another source has mentioned that to me. That they thought he died of alcoholism, or that that was the bottom line under the official cause of death listed.

Fred Wachs: I kind of had that impression myself.--Although when we knew him, when he was around us I never saw him drunk. But he looked like a person that drank a lot--his face was always red and flushed. He used to smoke cigars a lot. Loved to hunt--bird hunt.

Interviewer: Did you ever see Cape Grant in the show ring with one of the champions of Spindletop Stables?

Dorothy Wachs: Uh-huh; Uh-huh. Marvelous! Simply marvelous! There are some good Saddlebred trainers and riders now, but he was outstanding! I thought he was great in the ring!

PASSIONS AND PREJUDICE

Interviewer: Everything I've found in my research indicates that he was in a class by himself--perhaps the greatest in the history of the Saddlebred industry.

Dorothy Wachs: Those horses he could make them do anything--ah--Wonderful! Absolutely wonderful! A good showman on a horse--Tom Moore is possibly maybe as good.

Fred Wachs: Miss Pansy died soon after she gave me that Llewellyn pup. I knew she was in ill health. I wrote her a letter to thank her. And I got a letter back from her. She died real soon after I got the pup.

Dorothy Wachs: Fred's father was back and forth many times to Beaumont to see Miss Pansy. You know, she wore a hair piece?

Interviewer: What?

Dorothy Wachs: A hair piece. She wore a hair piece. She did. Well, it wasn't all that much--curls--curls around the bottom--around the sides of her face, you know.

Fred Wachs: You know, I can remember Miss Pansy--just plain, you know, like a farmer's wife--just plain-spoken--

SELECT ORAL HISTORIES

Dorothy Wachs: Well, she was that shape--fat, quite a bit--and very plain. And hats, you know everybody wore hats then. And I always thought her hats were a little off balance. But her dresses were simple, simple.--Little prints usually. And I think she had a dressmaker in Kansas or someplace out West. I don't know, it seemed to me she either went out or this woman came to her and I thought, "Gosh, you're going to all that trouble, she ought to do a little better." We never discussed clothes though when we were together.

Interviewer: What would you talk about in conversations?

Dorothy Wachs: Mostly, we listened. Cape, now--Cape was a big talker, and Fred's father was a great talker.

Fred Wachs: She talked some about the oil fields. Uh, mentioned some of that.

Interviewer: Did she ever mention Frank Yount, her husband?

Fred Wachs: Mr. Yount?

Interviewer: Yes.

Dorothy Wachs: She always called him Mr. Yount.

299

PASSIONS AND PREJUDICE

Fred Wachs: Yes, she called him Mr. Yount.

Dorothy Wachs: She told me, now the jewelry, she said, "You know, I just don't care about jewelry at all but Mr. Yount enjoyed buying it for me" and she said "He had lots of things made for me." Then she described some black pearls. She said, "I have pearls of every color--pink, black,-"-ask Mildred or Kathryn. I always wondered about what became of all this.--Fred's father was down in Beaumont--and he said their house down there was dark and heavy. Spindletop here was beautiful when she had it.

Fred Wachs: I just love Spindletop Hall down there on the first level--

Dorothy Wachs: The basement. (Laughs.)

Fred Wachs: Basement. She had an old western bar. And she had a lot of--I'd say maybe five--original Remingtons. I just loved to go down there and look at those.

Dorothy Wachs: Strictly a western bar. Sort of rough. Like out of the movies. (Laughs.) But the other room down there was a little room, with little twinkle stars on the ceiling, and it was called the "New Orleans" room. And, you know, it was for parties. And she had white wrought-iron furniture down there. The whole thing was very magnificent. A magnificent house! I can't think of the architect's name--he was from Louisville--uh--

Interviewer: Mr. Hutchings.

Dorothy Wachs: That was it. Hutchings. (Laugh.) A lady we used to know, he was her brother. This was during the Depression. And I mean, things were tough. And uh, Hutchings was sitting in his office one day in his shirt sleeves, hot, dejected--non-busy--not doing anything. And Pansy walked in and he said, "Yes. Did you want something." And she said yes, yes, she did. She was interested in building a house. Well, he sat up a little straighter. He did not get up. And he said, "What kind of a house?" And she described how many acres she had here and would like something around so many millions or so. So Mr. Hutchings gets out of his chair, holds her chair, gets his coat on, and said--"Yes!" (Laughs.) You know, he had been pretty low. (Laughs.)

Fred Wachs: Did you ever see that big painting of the nude they have. I guess it's in the men's room out at Spindletop now.--Well, she gave that thing to Dad, and he didn't know what to do with it. He put it in a back room, on the back of the house at the farm.

Dorothy Wachs: Back in the servants' rooms. (Laughs.)

PASSIONS AND PREJUDICE

Fred Wachs: I think he was embarrassed to have it. So he finally just sent it back after the University took over. And so they put it in the men's room. It's a full-sized nude. Looks like it was painted in about 1880.

Dorothy Wachs: Very voluptuous.

Fred Wachs: I can't remember where it was when Miss Pansy was there.

Dorothy Wachs: I never saw it out there. I remember the Reynolds that was in the, uh, music room, Sir Joshua, an oil painting. I know I was mad for years--provoked--they had a painting of a barnyard scene. It's fine, except, Victorian period--hanging in the dining room--which is a very elegant room. This is, of course, after the University took it over, and I don't know who in the world was doing--they had a lot of fake flowers--just awful. Here in this pretty, gorgeous room, (laughs) with its yellow satin walls, this barnyard scene hanging over the fireplace.--Not while she was there, that was afterward, when the University took it over.--Nooo.

Fred Wachs: I remember in Cape's bedroom upstairs, he had a big nickelodeon--had red, yellow lights all around it, you know.

Dorothy Wachs: Well, Miss Pansy had a beautiful pink French powder-room--on the first floor--on the main floor. French antique pieces also in the hallway. And there was a tier table. And on the tier table, there were a couple of Mexican pottery things with "cactus" in them. Now that was when she was there. And I was awfully upset about that--because I thought--you know--it wasn't working for me. (Laughs.) The room was so perfect and done so beautifully, and they had these pottery donkeys with cactus growing in them. (Laughs.)

Fred Wachs: She always wanted to walk you around in all the rooms and show you. She knew where everything came from, exactly what it cost--

Dorothy Wachs: She traveled all over Europe, and I don't know who--who helped her do that house. You know, the carpet in the entrance way was specially designed to match the ceiling.

Fred Wachs: And in the dining room, living room--both those carpets matched the ceilings--not the one's that are there now--but the originals.

Interviewer: Do you recall seeing two Stradivari violins out there?

Dorothy Wachs: Yes, I do. In the music room.

Fred Wachs: She had a lot of handsome books, you know, leather backs and everything, in the library.--She gave my father--they raised pheasants on the farm--when they left the farm, moved, she sent all the pheasants out to our farm, and we turned them loose. (Chuckles.)

Dorothy Wachs: Pea fowls.

Fred Wachs: Pea fowls, too. But these were pheasants. Ring-necked pheasants.

Dorothy Wachs: A menagerie. She had sheep--fancy goats--Pomeranians. There was a whole room in Spindletop that was the Dog Room......--You know, we were married in '36, and Miss Pansy sent us a pair of sterling silver candelabras--three branch. And Cape Grant sent us a sterling platter with monogram on it. And some place along the line one of them said, well you shouldn't have done that, they needed a carving set. So we also got a carving set. I don't know which one. But I think she said, well--send them one. She gave Fred's dad and mom an incredible Lazy Susan.

Fred Wachs: She called Dad for everything, seemed like. I remember he kind of laughed about it. She talked to him like he was an attorney. He said, now I'm not an attorney, Miss Pansy. But he did a lot of things for her that an attorney would do.

SELECT ORAL HISTORIES

Interviewer: Now he, Fred Wachs, Sr., was instrumental in the Spindletop sale to the University.

Fred Wachs: He handled it--totally handled it. Oh, she totally turned it over to him--to handle the sale. I know he did all the dealing with the Bishop, with Harold Lloyd when he came down here to look over the land, and with the University. Oh, I tell you who was President of the University of Kentucky then--Frank Dickey was President then. Of course, he was tickled to death--. And then they almost had the thing as a gift. They almost gave it to the University.

Dorothy Wachs: It was less than a million dollars exchanged.

Interviewer: $850,000.

Dorothy Wachs: Is that all?--That's it?! Well, that's just a token. Well, that's a big place!----Oh, I'll tell you something. That place had white fences-- white plank. They had them painted one year. The paint came from Sears and Roebuck. And it peeled off. And Miss Pansy was just furious--just awfully provoked. And that's when they put up the chain link fence around the outside--around the borders of the farm. Doing away with that white plank fence! (Laughs.)

PASSIONS AND PREJUDICE

Fred Wachs: Seems like when we went out there we always went in not at the main entrance gate that leads to the house now, but there was another gate further down toward the Iron Works Pike--I mean, toward Newtown Pike.

Dorothy Wachs: Oh,--treacherous thing. It had spikes this long--(indicates long metal spikes with hands)--that if you drove into it from the road when it was locked--your tires were gone. (Laughs.)

Fred Wachs: They always kept that locked. And my dad had a key. We'd go out there, and he'd unlock the gate, and we'd come in around at the back of the house somewhere. We didn't come in the front driveway. Matter of fact, I don't remember ever coming in the front way until after the University got it.

Dorothy Wachs: But the whole perimeter of the farm was chain-link fence.--Which I despised. Oh, I guess, I didn't exactly despise it. (Laughs.) --I think she must have loved those horses--all those horses--loved the animals.

Fred Wachs: I can't remember all the animals. But I remember behind that house, she had all kinds of different pens with different birds, and peacocks, and pheasants. And kind of odd farm animals--different kinds of sheep, and, uh,--

SELECT ORAL HISTORIES

Dorothy Wachs: They were all purebred, whatever they were.

Fred Wachs: But she didn't get interested in the bird dogs until way late in life. And I really think it was to just try to beat Cape at his own game. But I think that was her idea.--She liked to tell stories about the oil camps. I heard she was a cook at one time for them.--She had a real plain talk--her accent and everything was strictly, plain Texas.

Dorothy Wachs: I think what you saw was what you got with Miss Pansy. I don't think she even wanted to fool with anyone, uh, who wasn't who they were--(Laughs.)

SELECT ORAL HISTORIES

Mr. George Stanhope Wiedemann
Oral History Interview
July 25, 1994
Lexington, Kentucky

Interviewer: Linda Light
Subject: Pansy Yount of Spindletop

"Hope" Wiedemann: Purcell's Department Store of Lexington, Kentucky, was Lexington's most famous department store in the 1950's. At Christmas time, the store decorations were outstanding, because in our medium-sized city the best trade was induced by having a super decorated store. We had earned a great reputation over the years. The credit at the time was due to the efforts of our excellent display man, Paul Freeman, who was also responsible for the display windows.

The store decorations, which were in place by the day after Thanksgiving, did not escape the notice of Mrs. Miles Frank Yount of Spindletop Hall. She called me--I was the Store Manager at the time--to request that she could engage our display man to decorate Spindletop Hall for Christmas. I asked Paul whether or not he wished to tackle the job, in addition to his other duties? He said he would appreciate the opportunity. Mrs. Yount was a

PASSIONS AND PREJUDICE

friend and an excellent customer of the store and I wasn't about to refuse her request.

Paul did such a good job at Spindletop that Mrs. Yount wanted him to go to Beaumont, Texas, and decorate her house there. He went with my blessing. Upon his return he said he had never been treated so royally by anyone in his life and he and Mrs. Yount became good friends. Paul continued to decorate her houses until she moved back to Beaumont.

When Mrs. Yount would appear at the store to shop, I would be alerted by the sales clerks. I would always greet her and thank her for coming and ask if I could be of service.

One time, she was having a barbecue on the palatial patio at Spindletop for twenty-four guests. She wanted twenty-four pancake spatulas "because each guest should be equipped to cook his or her own hamburger the way they liked it." We took notice of her with that kind of order!

One thing--I think Miss Pansy thought all the society stuff and social rules here were a bunch of "crap."

SELECT ORAL HISTORIES

Interview with Mr. and Mrs. Jack Grant (Jack Grant is the brother of legendary Saddlebred trainer, Cape Grant)

January 1, 1996
At their home outside of Texarkana, Texas

Interviewer: Linda Light

Subject: Mr. Grant's brother, Cape Grant; Pansy Yount; Spindletop

Jack Grant: Frank Grant, Cape's youngest boy, you met him, didn't you?

Interviewer: Yes. I did.

Jack Grant: And most of the others. And you know, they are the only ones other than me. Cape's boys and me, are the only one's still living, I guess. I'm sure this doesn't hurt them any, and it doesn't hurt me any. My wife and I were discussing this today. Am I being recorded now?

Interviewer: Yes. Is that all right? I want to use this to round out the new edition of the book.

PASSIONS AND PREJUDICE

Jack Grant: Why, that's all right. And we were talkin' about this situation and how it came about. And she mentioned--she said, you know, this type of thing that happened to Cape happens every day. Some man with a secretary--a younger secretary--workin' with him everyday--she's tryin' to please him. He's tryin' to please his spouse and it just--like I say--it's just somethin' that occurs. Come in and set down with us, Kathleen.

Mrs. Grant (Kathleen): I just wanted to find out if you'd like coffee or tea.

Interviewer: Coffee'd be great--if ya'll are goin' to have some.

Mrs. Grant: Oh, we drink coffee just 'bout all the time. (All laugh.)

Interviewer: O.K., then. Thank you. Well, you know, I think that makes it a very modern story. Without Cape, it wouldn't be the story it is. It's very modern and people can identify.

Jack Grant: Well, it was just one of those things. And she was his secretary. And it just developed, I guess.

Interviewer: And this was Mrs., uh, I just got through reading it in the divorce papers. She was down in Beaumont.

SELECT ORAL HISTORIES

Jack Grant: Odom, I believe.

Interviewer: Is that what happened then between him and Pansy? I mean, she sort of caught him, or knew....

Jack Grant: She caught him with her in the office, I believe, that's what brought it all about. And of course, at the time, we said that old trite sayin' that, uh, "There's nothing like a woman scorned," you know. I mean, (laughing) vengeance comes in all sizes--you know, I'm sure--to get even. Of course, I can't blame her; and I can't say that I blame him. My mother was real good to me and to our family. And the incident about Cape telling, uh, his son that if he got married that he'd cut him off financially. This happened, but Cape didn't have a tremendous amount of money. He spent quite a bit, but he actually didn't have a whole lot.

Interviewer: He told his son that?

Jack Grant: No, he didn't. I mean, I know for a fact that he didn't. He didn't. He didn't have a tremendous amount of money anyway. And Cape's theory was, if his son married, there'd be a child. There'd be this and that. Perhaps it'd cause the boy to drop out of school and everything. He just put it on ice, "Now, if you get married, you're cut off. No more funds."

Interviewer: Which son was that?

PASSIONS AND PREJUDICE

Jack Grant: Silas, the doctor.

Interviewer: Oh, I didn't know that. It's impossible to know everything.

Jack Grant: (Chuckling) Yeah, I was surprised that you found out half as much as you did.

Interviewer: But I've often wondered what happened. I know that Silas was the son the family wanted least disturbed. And that's possibly why.

Jack Grant: Well, too, Silas--he, he has Parkinson's disease now. But, he and Cape and his family were on good terms when Cape died.

Cape went to Silas' hospital in Hillsboro for an examination. Cape had a heart condition. And while he was there, Cape had gone back to smoking cigarettes. He had quite for a long time. And Cape--the way it was told to me--Cape dropped a cigarette on the floor and reached over to get it. And that's when--there was this coronary thing. And he lived for a few days after that. They operated on him. They couldn't do anything for him though. He lived three or four days after that. But he was back on good terms with his wife, and with Silas. And Si had been treatin' him as his doctor. Oh, I don't know the whole thing. I've heard things. I've heard Cape say though that one

SELECT ORAL HISTORIES

reason he married Pansy was that all his life he had wanted the best stables, and saddle horses money could buy. And that's how he could get it. Then I also heard they were in a wreck down in Liberty--down in that area of South Texas. I think they were kinda' on the outs with Mildred at that time.

Interviewer: Why were they on the outs with Mildred?

Jack Grant: I don't know. I just don't know. But they were for some reason. And, uh, Pansy more or less said, well, that if anything happens to me, then Cape may be cut completely out and not get anything from my estate. And--so that's before they were married--so that's another reason why I heard they married. But I don't know.

Interviewer: So that he would definitely get something from her. Because he did so much with the horses for Pansy.

Jack Grant: Yeah, not only that, he did everything, Linda. The staff, the people who worked in the house, plus all the farm people that he had working on the farm and all. It was a tremendous working responsibility. He earned every dime that he got, believe me. When they were in Beaumont--which they were six months out of every year--he'd go to Dallas every day to see about her investments, stocks and all. He did all that.

PASSIONS AND PREJUDICE

I get amused at some of the things that happened. She was quite entertaining at times. Uh, she told me one time, says: "We were at a horse show and," she said, "this lady came to me and says 'Cape? Cape? Why do you call him Cape?" She said, "I guess it's because his mother named him that." (Laughs.)

And we were goin' one time to Louisville to a horse show. And we were in this limousine and a chauffeur drivin' and everything. And, uh, I don't know if he was drivin' or not--but he was with us. And I think that he went so that after the horse show he could bring us back to the farm and they were gonna stay over. He was an ol' Georgia cracker. What he was really was a fishin' guide. But they liked him well enough that they brought him to Kentucky as their chauffeur. So we were drivin' along there, and so he said--he went up that big hill goin' to Frankfort and he said, "This sure would be a good place to stop and throw some rocks. And Pansy said, "Cape, let him stop and throw some rocks if he wants to." (Laughs) But there were some funny things that happened up there. We had fun. But there was a lot of work involved.

Interviewer: What kind of person was she? What was your impression of Pansy Yount?

Jack Grant: She read a lot. And I recall one time she kept askin' me, "Well, what are you readin'?" And at that time, I said, "Well, I'm tryin' to read the Holy Bible from beginnin' to end. I guess I'll just slow read it." And then Pansy goes out and buys me one of these easier versions to read, and gave it to me. But she--we always got a Christmas bonus--everybody that worked at the farm did. And she'd give us money to buy presents for my two sons. She was very generous in every way. And there was a ritual--coffee at 10:00 a.m. and 5:00 p.m.

Mrs. Grant: It stood up and said "coffee," too! (Laughing) It was strong. (Laughter)

Interviewer: How long were you in Kentucky on the farm?

Jack Grant: It so happened that I was the farm manager, but, uh, as time progressed, they got less and less stuff so it finally wound up to be just a matter of maintenance. But we kept the leaves raked up and the yards mowed and everything was really in tip top shape. But we finally got down to where we just had a few saddle horses and very few cows and maybe some sheep, and that sort of thing. But it just kept getting less and less every year. And then one time she talked about moving all the people back to Texas--the whole farm personnel--but things didn't work out.

PASSIONS AND PREJUDICE

Interviewer: Did you notice any differences between up there in Kentucky and here in Texas during those years, Mrs. Grant?

Mrs. Grant: Oh, I think we did for the simple reason that we had not been around other states. I mean, Jack had; but I had not.

Interviewer: How did Kentucky strike you during that time? It must have been fabulous.

Mrs. Grant: There were so many things that were different. It was enlightening--an eye-opener--and mind-bogglin' and--

Interviewer: You knew Pansy, too, didn't you, Mrs. Grant? What was your impression of her?

Mrs. Grant: Oh, I liked her. (Laughing) To me, she was just down to earth and so--she knew what she wanted and she--she exacted that from the people who worked for her. One time she decided that I should make jelly, and the gardener would bring over all this stuff for me to use. (Chuckling) He brought over so much cabbage at one point--I was running it down this thing (laughing)--and rather than fibbing about it--I said that that was my way of makin' sauerkraut. (Laughs heartily.) Then I know one time I asked the gardener--he had brought me so much squash, you know, and a family of four--we thought we'd

SELECT ORAL HISTORIES

turn into a squash. So I finally said, "Mr. so-and-so, how does your wife prepare squash?" Well, he didn't know but said he'd let me know. And he, a couple of days later, I spoke to him again and he said you know my wife told me to tell you she fries squash. And I said, "Well, how does she fry squash?" Cause my mother had never fried squash. And he said, "Just like fried cucumbers." (Laughs) When I told my mother, she thought that was so funny, since she had never heard of fried cucumbers.

Interviewer: How did people up there in Kentucky receive Pansy, do you think?

Jack Grant: Well, she didn't have many friends. I wasn't aware of them--they may have had 'em--but, if they did, I wasn't aware of them. Well, you're from Texas. You know, this hard boot attitude. I love the people up there in Kentucky, but if you don't have a pedigree and haven't been there all your life, why--you're not accepted readily--I'll just put it that a way.

Mrs. Grant: It may still be that way, but people now have missed out on the real Bluegrass. It was the heart and soul of the horse industry really way back then.

Jack Grant: When we where there, there was a saying that you either work with horses or you were a horse. There was hardly any middle

PASSIONS AND PREJUDICE

class. I mean it was all pretty high. Of course, we were middle class or little lower. But there was not very much. But of course, IBM came in there and different industries--Square D--they sort of came in there and established a middle class that didn't exist there in Lexington before then.

Interviewer: How about the horses they had? You rode for a while, didn't you?

Jack Grant: Well, yeah--but I didn't ride after I went up there except for one time after the farm was practically closed down and they had a horse they were exercising. They had a man on the farm. His name was Billy Macalvein and he was a very good horseman and good friend of Cape's and he'd been their other trainer during the time I was there. And Mac was good and he had, uh, been when he was twenty-one years old, as I understood it, he had worked for Mr. Adair in Atlanta. And Mr. Adair owned the Coca Cola Company. And Mac was twenty-one. And back in those days, money was money. And I think Mac made a thousand dollars a month, even then, way back then. But Mac was still there at Spindletop Farm in Lexington and he had this one horse he was still riding and exercising. And Mac was the sort of guy that if he could get out of doing something--well, he got me back to riding this horse he had. But from the time I was a youngster, I didn't have much to do as far as riding horses.

SELECT ORAL HISTORIES

Interviewer: How did that happen that Cape rode more, and you didn't?

Jack Grant: Well, that was the thing that Cape knew--and he knew well, and he loved it. So he did it. I liked horses and had my father lived, I would have probably wound up in the show horse business. But he died when I was nine, so....

Interviewer: And it was your father that taught Cape horses?

Jack Grant: Yes.

Interviewer: And your father was in show horses himself?

Jack Grant: Yeah, he was. My family--I do fairly well--but my dad and my brother, uh, wrote.

Mrs. Grant: Jack has pretty handwriting.

Interviewer: Artistic.

Jack Grant: (Showing Interviewer a copy of a letter written by his dad.) And this is the way my father wrote. And he could write--the letters that he wrote would be shaded naturally. Anyway, this is a letter my father wrote. See how the writing was in his letters and Cape's. Artistic.

PASSIONS AND PREJUDICE

Interviewer: Do you know what Cape's first wife did after they split up?

Jack Grant: I don't know. I know she moved out of the apartments there. I think she lived in some apartments owned by the Younts, and she moved to San Antonio with her sons. I think she must have got some kind of job down there. But I don't know just what it was.

Mrs. Grant: Oh, I know what I wanted to tell you. In your book, the carriage in front of the house, that's Macalvein on that carriage. I believe his wife and Stansbery worked in the house.

Jack Grant: One of the stories I always enjoyed--I wasn't there when it happened. Many, many people heated with fuel oil. So you could imagine they'd have a pretty good reservoir to heat the mansion in Lexington. Well, one time the reservoir started leaking, but they didn't know it. And there was a spring down, not too far from the house. And they went down there and there was oil all over and they took Pansy down there to see it; and she put her hands up and said, "By God, Cape, we're rich again!" It was that stuff goin' through that limestone running down that....(Laughing)

Mrs. Grant: She just really enjoyed just plain, simple things. She'd just laugh and enjoy things. And then she'd turn around and tell it so

SELECT ORAL HISTORIES

interesting and funny, whatever little instance happened.

Interviewer: So she had a good sense of humor, did she?

Jack Grant: She did. She sure did. She had several dogs. And she had a little dachshund she called Jimmy. And, uh, she was talkin' one time about Jimmy catchin' this mole. And you know how a mole looks. And Jimmy had him in his mouth. And she said, "And there was Jimmy with that mole, with the mole's hands sticking out his mouth." (Laughs heartily.) (Focusing again on the letters of Cape and his father)

Interviewer: (Looking at Cape's letter) He must have been incredibly well-educated with that handwriting.

Jack Grant: Self-educated. Cape never went to any college.

Mrs. Grant: But you'd never know it.

Jack Grant: No. No, you wouldn't. And he was very gracious. I mean, helping a lady pull out the chair for her. And standing whenever a lady got up to go. And all this kind of stuff.
Mrs. Grant: The mannerisms of Cape and Pansy were different. I mean, his was more

PASSIONS AND PREJUDICE

polished from the standpoint of--probably his years of training show horses.

Interviewer: More refined then.

Jack Grant: Well, I think, too, the clientele he dealt with. Not everybody can afford to have a stake in a show horse or even two or three. That's pretty expensive. You know, they're pretty affluent types. So he more or less learned by being on the scene.

Mrs. Grant: I'll tell you, his mother and father, from all I knew, were just genteel. He grew up with that.

Interviewer: Someone told me she was a school teacher. Is that true?

Jack Grant: No. She did attend some--you know, at Denton they used to have a CIA, a College of Industrial Arts, like homemaking and stuff like that. And I think she attended school there. But she never taught. But Cape's and my mother was the type of person that everyday was gonna be a better day than today. And I think that's a wonderful attitude. I think she felt there was something good gone happen.

Interviewer: Did Cape pick up that optimism, do you think? You obviously did.

SELECT ORAL HISTORIES

Jack Grant: I suppose he did. His looks and coloring and all were more like my mother now. The rest of us were blond and more like our dad.

Interviewer: People have described Cape to me as tall and dark.

Jack Grant: Yeah, he was.

Mrs. Grant: He looked to me--he almost looked like George Raft, the actor.

Jack Grant: You know, I've had some of the most ridiculous questions posed to me about that mansion in Lexington.

Interviewer: Like what?

Jack Grant: Like I had once one of those people in Kentucky ask me if she didn't have a room somewhere in the mansion papered with ten-dollar bills or somethin' like that. You know, ridiculous. But they really had heard that and believed it.

Mrs. Grant: But her tastes were far and above the average.

Jack Grant: Oh my goodness, yes. I think you talked to one of the original painters--he hand-painted these borders on the walls and decorated it and all.

PASSIONS AND PREJUDICE

Interviewer: Did she personally pick out the designs and the way things were decorated?

Jack Grant: I'm sure she did. She had, uh-- she had very good taste and everything.

Mrs. Grant: Another thing, she liked to read English novels, too. So she picked up a lot of ideas through her reading.

Interviewer: Do you know what her favorite book was?

Mrs. Grant: She had a collection of books. She even received a commendation from the Smithsonian Institute on her collection of Sherlock Holmes. Of course, it just constituted one section of her library.

Interviewer: So she really was well-read?

Both Jack and Kathleen Grant say almost in unison: <u>Oh, my yes</u>. She sure was.

Mrs. Grant: If you could hear her talk, she sounded closer to East Texas and Arkansas than I do. (Laughs) But what was in her head was far different than her voice would let on. But that really hurt her. Her face and her features belied her knowledge.

SELECT ORAL HISTORIES

Interviewer: What did Pansy look like? Can you describe her?

Mrs. Grant: (Looking at Jack:) Should I say it? (Laughing)

Jack Grant: Go ahead.

Mrs. Grant: Minnie Pearl. (Laughter) When Pansy laughed, she made you want to join her. Her face just broke into laughter, and it was just beauty. Just plain beautiful when she enjoyed something.

Interviewer: I'm afraid I've only gotten the Kentucky slant on Pansy--you know, the rooms papered with ten-dollar bills, thus far. So I'm glad to hear the Texas side.

Mrs. Grant: And another one is Nola, Cape's wife. Nola once again is a person with a wonderful sense of humor. She's a very beautiful person. I've never been around her that much, but the times we have been around her, she was very beautiful--especially for her age.

Jack Grant: And as far as Cape's drinking was concerned--he never drank anything when we were at Spindletop. I may have told you this, but people like to think there was somethin'. And Pansy would say, "Maybe these two would like somethin' to drink." "Aw, they don't need anything

PASSIONS AND PREJUDICE

to drink," he'd say. (Laughing) Kinda rubbed me the wrong way, you know. One of his favorite sayings was: "Drinkin' is a full-time job. I know. I used to do it." And another thing was--I think I told you one time--that he had this bar--why it's down there with Nola now--had this chrome bar that you could wheel around. You could see it when you were goin' into that study. But anyway, he had that and we'd talk about drinkin' and everything. He'd talk about how he hadn't had a drink since so-and-so. And he'd say, "That whiskey is twenty-five years old." Well, maybe it was ten years old, but he'd had it fifteen or something. And we'd always kid him, "Well, that stuff won't age in glass, you know." (Laughs) Well, anyway, he didn't drink any while we were up there now. When this divorce with Pansy came about, he started drinking again. But he definitely wasn't an alcoholic. But now I did hear him say though that when he was showing horses--now that's before he went to work for the Younts--that he used to drink a quart of whiskey--I think he said a quart of whiskey--every night and get up at four o'clock in the morning and train horses. But an alcoholic to me is someone who has to have to drink. He never did have that kind.

Interviewer: It did take a toll on him though when he and Pansy got the divorce.

Jack Grant: Yes. Yes, I'm sure it did. I don't know. But I'm sure that's--that's one of the things that caused him to drink again. But this bird dog

SELECT ORAL HISTORIES

thing you mentioned in your book is true. Cape imported bird dogs from Llewellyn--these Llewellyn Setters. Now a Llewellyn Setter is much like an English Setter, but they're bred by this man named Llewellyn over in England. And they have been around a long time. And Cape told a story, said "When I was a young man," he said, uh, "I wanted a bird dog." And Dad said, "I know a man in Texas"-- and that's when they were still in Oklahoma. And he said, "I know a man in Texas that has some fine bird dogs." He said, "I'll get you one. So what do you want, a pointer or a setter." Cape said, "Well, really I'd like to have a setter--I mean, a pointer." So, Dad said, "Well, you get a pointer and I'll get a setter." So they had the man send the dogs in. "Well," Cape said, "That dog of mine was a liver-spotted pointer, and I loved that dog. Taught him to retrieve a ball and do all this and everything." But he said, "I just liked to hunt more with my dad's setter. I got so's I'd take him and leave this other dog home." Cape said, "Now that dog would do everything that I wanted her to." Said, "I named her Rarity's Queen." Said, "She'd do just fine. She'd find the birds and retrieve them, but," said, "one time I shot a hawk. And she went to get this hawk, and it grabbed her with its claws right in the face." And said, "She wouldn't retrieve for a long time." But said, "Finally, I got her back to it." But that was his first experience with Llewellyn Setters. But he imported some while Mr. Llewellyn was still living and Cape bought some from him. And that was just about the time before this divorce thing. So it was pretty competitive--so Pansy did get a bunch of dogs. Cape and Pansy were very competitive.

Mrs. Grant: Speaking of competitive, did you tell Linda about (chuckling a little)--about the servants--

Jack Grant: Ohhh. They had servants and most of them were from Texas that came up. Blacks. And, uh, be there and everything. And they had one there--and, I don't know--I saw Cape one morning, and he was pretty irate. And I said, "What's the matter today?" And Cape said, "Well, it's ol' Ira." He said, "Ira's been up there and he's been around here for two or three years and I thought he was workin' for Pansy and she thought he was workin' for me." (Laughing) Well, you see, he'd have his servants and she'd have hers and there were so many--they got to checkin' up one day and found that Ol' Ira, as they called him, was neither one of them's servant. (Laughing)

Interviewer: Why didn't they have servants from Kentucky?

Jack Grant: Well, I don't know. She had some cooks from Kentucky, I think. I don't know whether they couldn't get 'em or whether--I think when they were building that mansion up there, I know they had a strike and there may have been something to do with that--that she thought that

SELECT ORAL HISTORIES

was what all these Kentucky people were like or somethin' like that. But anyway, they had a strike and some of them that worked on the farm, they would slip in the back way and come to their quarters, but Cape told 'em. He says, "Now I don't have anything to do with that house. Whatever ya'll do about that is no problem with me." And he says, "If you interfere with this stable or training these horses," he said, "there's gonna be some trouble--there's gonna be a lot of trouble." So they never did bother him, and they let his people come into town.

Interviewer: Were you at Spindletop when Beau Peavine and Chief of Spindletop were there?

Jack Grant: Yeah, but they were about pension age by then. I remember when ol' Beau died and, of course, you know the Roxie Highland statue--they had thought about making one for Beau Peavine, but somehow--

Interviewer: Do you know where they buried Chief of Spindletop?

Jack Grant: I think it was right there by the flag pole where we buried Beau Peavine--right next to that house where we lived.

Interviewer: You think that's where Beau Peavine is buried?

PASSIONS AND PREJUDICE

Jack Grant: Beau Peavine is.

Interviewer: How about Chief of Spindletop?

Jack Grant: Chief is buried up there by the barn.

Mrs. Grant: He's buried right up there on that hill.

Jack Grant: Yeah, right by the barn.

Interviewer: What hill?

Mrs. Grant: Right there where the carriage house and the training barn is. Right behind the training barn. That's where Chief of Spindletop is buried.

Jack Grant: There was a marker there at one time. Somebody may have--

Mrs. Grant: With his name and his years.

Interviewer: Really. So somebody may have taken it.

Jack Grant: Uh, Ed Manion. Cape had a stables in Tulsa, Oklahoma. And I was about--oh, I couldn't have been over eleven or twelve years old--uh, Mr. Manion, Eddie Manion, would come

SELECT ORAL HISTORIES

there and ride. And you can get the era when I tell you what I saw him come out and do. He came out there and had a ukulele. And this was back there. And he played and sang a song for us out there. But he, they, kept a horse out there, and he rode it. So, uh, Cape introduced him to Mildred later on--Mildred Yount--and I think the first time they went out. But, of course, Eddie Manion was prematurely gray so Mildred came to Cape and said, "Mr. Grant, that man is gray-headed. And Cape said, "Why, hell, I didn't mean for you to marry him or anything. I just wanted you to go out with him." (Laughs.) Of course, they did end up getting married, you know. And Mr. Manion, Eddie Manion's father was the President of the Pipe Line for Sinclair Oil Company. But I knew Eddie a long time before I knew the name Yount or anything like that.

Look here, at this. It's an article on Cape and the way he trained horses. Here's another thing I found on the mausoleum Pansy built.

Interviewer: Such history. It would be a shame to lose it.

Jack Grant: Have you seen the Mildred Building there in Beaumont? At 460 Mariposa?

Interviewer: Yes, sir.

Jack Grant: Well, it cost a million and one-hundred and ten dollars. Of course, the

333

PASSIONS AND PREJUDICE

apartments go with it, too. And they say that Mr. Yount personally witnessed every shovel full of sand that went into that building. I mean, they were right on the job, watching everything when it was done.

Mrs. Grant: Here's a gourmet cook book Pansy gave me.

Interviewer: She had a beautiful handwriting, didn't she? I also interviewed Katherine, the granddaughter of Pansy.

Jack Grant: They were just little kids then. I know those kids were up at the farm, but they were never there when I was there. And they'd see the kids and talk to 'em when they came to Texas. And I know that at one time--at first--I worked for Cape and Pansy, uh, at Beaumont, in the office down there for a while. But I didn't like the working inside, so that's the reason I took another job. But they would--the man that was in charge of their office down there said, "Well, Mildred gave each one of the kids $125,000 this year." So that's about all I'd hear of them. I never did see them or anything. After they had split up that estate, well, Pansy's end of it didn't have anything to do with Mildred's. So I didn't know what was goin' on.

Interviewer: In Kentucky, they tell that Pansy was fat. Was she?

SELECT ORAL HISTORIES

Jack Grant: Pansy was stocky, heavy set, but, no, I wouldn't call her fat. No. And she had beautiful hands.

Mrs. Grant: And she used those hands very effectively.

Interviewer: I did an interview with someone in Kentucky that said she had hands like a farmer.

Mrs. Grant: No. I thought she had pretty hands. Had they been younger hands. She used them so effectively that I just had an appreciation for hands after that.

Interviewer: I'm glad to include this interview in the next edition of the book--another side of the story.

Jack Grant: On the cover of your book is a beautiful picture of Pansy. I think it's great. As I remember, she didn't like to be photographed. She wore a lot of black. One time, we went out, and she had on the most beautiful linen dress I ever saw. And we admired it and said how beautiful it was. And later she gave it to my wife.

Interviewer: Do you know anything about Pansy's background? There are various stories. I think she had some brothers and sisters that died early.

Jack Grant: Yeah, she had a sister that lived in Beaumont, I know. She came up to Kentucky once, and I remember she looked around Spindletop Hall and said, "My word, Pansy, I don't see how you find yourself around here in this house, it's so big." She had a brother, too, I know. But she didn't cater so much to him. Pansy was Catholic, you know. I remember she told Fred Wachs, Sr., once, said, "Hey, I just delivered a turkey to Father McKenna--gave him a turkey. He gave me another medal." Fred said, "Oh, he's always giving out medals." Pansy said, "I wish he would give me something else." Fred Wachs, Sr., took Pansy's side in the divorce. Fred Wachs, Sr. was the type of person that anything they mentioned that they'd want, Fred Wachs could get it for them. No matter what it was, Fred had the connections to get it. But about Pansy being Catholic--first of all, though, I thought I'd call her "Miss" Pansy. But she told me, "Just call me 'Aunt Pansy.' I'm your Aunt." I thought, well if she's got that much money, you'd call her by anything she wants to be called. (Laughs) But about being Catholic--when I was in the service, Pansy sent me a Catholic medal and it said on the back of it: "Jack Grant. I'm a Catholic. Call a priest." But I wasn't Catholic. Right now we go to the Baptist Church. That's what my wife is, Baptist. But Pansy sent me this Catholic medal and said, "You be sure and wear that, Jack." Her idea was that I might be goin' to Italy in the war and if I did get wounded or somethin' over there, I'd at least get top grade service. (Laughs)

And horse stories. I remember some good horse stories that Cape had a part in. There were two horses, one Shamrock, the other Denmark-- their names were longer. Anyway, they were full brothers. Outstanding Saddlebred horses. He wanted to buy one of these horses. So, he said, "I stayed on that train with the guy that owned those horses and talked to him and talked to him. Even got little tears in my eyes about that horse, I wanted him so bad."

Another horse story, when Cape bought Roxie Highland, Swede McGuinness was at the farm. Cape one day went out to the barn and said, "Who's takin' care of this mare?"

Swede said, "I am, Mr. Grant."

"Well, "said Cape, "I want you to make sure she's taken care of and everything until we send for her. Here's twenty dollars, get you some breakfast."

Swede said, "Well, that's just fine, Mr. Grant. But what about my lunch?" (Laughing) So I don't know if he gave him some more money or what.

PASSIONS AND PREJUDICE

Then another story I heard. Part of this I know is true because they had this Duesenberg car. They bought it at the World's Fair in Chicago when they had horses up there. It was on display up there. They weren't married then, I don't guess. Well, anyway, when World War II came along they gave that car to the scrap guy, because of Germany. They were fightin' against Germany, and they didn't want anything to do with anything German-made at that time. So they had taken the car to the scrap guy. A few days later that Duesenberg showed up on a dealer's lot. And they said Cape found out and took an ax down there and just chopped that thing all to pieces.

Mrs. Grant: But he would do that. And you could almost hear those two talking about that car, getting rid of it. They were just that patriotic. Oh, Cape had a big, beautiful juke box in his bedroom at Spindletop. He was very artistic.

Jack Grant: Oh, yeah, Cape could sing and play the guitar and banjo. And he liked to tell stories. He told one once, that he had bought this young man who milked cows at the farm, a good, warm, long coat so he wouldn't get so cold while milkin' the cows. Well, the young man went out and traded that coat for a pistol. Then, it came a "blue norther." Being from Texas, you know what that is, meant it got *real cold*. So it came the next morning, and the father of the young man was tryin' to get him up. It was just a-freezin', you know. The father called up to the young man's room and said, "Come on, John. Time to go to the barn and milk the cows." It was just a-freezin' that morning. Said, "Strap on your six-shooter, son, and go milk the cows."

This divorce thing though between Pansy and Cape, ate at them both, I think.

Mrs. Grant: I think it did, too.

Jack Grant: After this divorce thing, though, between Pansy and Cape, I was pretty despondent. I gave them pretty much the best years of my life. And, I think, I was forty-one years old, and it's kinda hard for somebody to go out and get a job at forty-one and there's not a big demand for farm managers, especially down here in Texas. It didn't pay very much at that time. So, I was kind of left out in the cold. Now, she did give me some severance pay, but I'm sure that Cape was sincere. He once told me, "You know, Jack, you have a wonderful future here at Spindletop." I thought so, and I'm sure that he did at the time, but it didn't prove to be that way. So there I was with two kids that went to the Sayre School up in Kentucky. I came back down here and put them in public school, had to find a house, and relocate and all of that. Of course, we wanted to come back to Texas.

PASSIONS AND PREJUDICE

Interviewer: The divorce of Cape and Pansy had sort of a domino effect on your family.

Jack Grant: Yes. But God has blessed us, and I'm seventy-seven years old. Anyhow, we've been fortunate. Very, very fortunate. I don't regret it. The boys still have memories of the farm and all. As far as what all happened between Cape and Pansy, as the old saying goes, "There's a little good in the worst of us, and a little bad in the best of us."

A Dr. Burns who knew Cape quite well told me that Cape once said, "When I die, Doc, I hope people talk about me. I don't care what they say about me, just so long as they talk about me." (Pausing, then smiling a little as he imitates Cape's voice) "Talk about me," he'd say. "Talk about me."

SELECT ORAL HISTORIES

EXCERPTS FROM A LETTER FROM:
MR. MARION J. ANDERSON
HOPE, INDIANA
DATED FEBRUARY 11, 1997

"I was born on the Shoshone Horse Farm where my father, W. R. Anderson, was employed at the time of my birth. My father showed me the little house where I was born. It was torn down when the Spindletop Hall building was erected. After Mrs. Yount bought the farm, my father worked as night watchman on the Spindletop Farm. He checked all barns and gates on the farm. He also checked the house where Mrs. Yount lived at Ironworks and Newtown Pikes. When I was about ten or eleven years of age, my dad would let me go with him to make his watchman rounds. My dad told me many good things about Spindletop.

Mr. Cape Grant was the manager and horse trainer of the farm. He was my dad's boss.

One thing I remember well was when our house burned on April 27, 1937, and I lost my baby sister in this fire. Mrs. Yount returned from Texas after a few days and when my dad was checking the house where she and Mr. Grant lived, Mrs. Yount came out and talked to Dad about the fire and loss of my sister. She handed my dad a piece

PASSIONS AND PREJUDICE

of paper, and he put it in his pocket. Next day, Dad took my brothers and sisters to Lexington and bought new clothing and shoes for each one of us. Dad also told me that Mrs. Yount bought the burial plot at Hillcrest Cemetery where Irene, my little sister, was buried. It is things like this that we don't forget.

On March 10, 1938, my youngest brother was born at a hospital in Lexington. My dad named my baby brother after Cape Grant. His name, Cape Anderson. He now lives in Lexington.

Mr. father told me that Mr. Grant paid the hospital bill when little Cape Anderson was born. Dad thought Mr. Grant was a very great man and that is why he named Cape after him.

A Mr. Curry was the manager of the dairy at Spindletop Farm.

Marion J. Anderson
Hope, Indiana

SELECT ORAL HISTORIES

**Interview with Saddlebred show horse trainer Art Ledbetter, Henderson, Kentucky
January, 1997**

Interviewer: Linda Light

RE: The Owen Hailey Incident

Yes, I knew Owen Hailey. Owen was disfigured for life in a car accident with Cape Grant. Hailey was one of the all-time great Saddlebred trainers--just a great, great horseman--and he was working then with Cape at Spindletop Farms in Lexington.

But they took Owen to the hospital when that happened. His face was cut up pretty bad. Anyway, they took him to the hospital and things were goin' along pretty good. They had to have surgery and everything on his face. And Pansy Yount was paying everything on it. And some fella came in that Hailey knew and was talkin' to him a little while, and he said, "Owen, you gonna get anything out of this?"

Owen said, "Well, you know, I don't know. Mrs. Yount is paying the hospital and takin' care of everything."

PASSIONS AND PREJUDICE

The man said, "You know, you better get yourself a lawyer and sue that woman."

So Owen told me, "You know, I did see a lawyer about suing her. I never got anything. And she never paid another dime on it."

So, you see, it was one of those things where somebody else got Owen in trouble. Owen told me that if he had just went on with the way it was, that lady, Mrs. Yount, would have paid everything. "If there had had to be any other surgery on my face," he said, "it wouldn't have cost me anything. But you see, Art," he said, "when I got this lawyer, she had a lawyer and she--she never paid anymore. Just got out of it all. I think it really got away with her. That's what I got for listening to someone else when Mrs. Yount was gonna pay for everything."

Owen Hailey was one of the top showman and horseman that I've ever known. He was later inducted into the Horseman's Hall of Fame. He came to my place when I was at Lake Zurich. Brought his wife; that's when we visited, and he told me all of this. But Mrs. Yount would have taken care of everything, so that's one time he got himself in trouble by listening to somebody else.

APPENDIX IV

Core Documents from Spindletop Times

CORE DOCUMENTS

TABLE OF CONTENTS FOR CORE DOCUMENTS

A. **Court Records**

    1. W. C. Grant vs. Pansy M. Grant for Divorce 349
    2. Pansy Grant vs. Cape Grant: Countersuit for Divorce 357

B. **Newspaper Clippings**

    1. Stradivari Violins 370
    2. Miles Frank Yount Honored by Rotary in Beaumont 372
    3. Funeral of M. F. Yount Horse Show Goes On 376
    4. Yount Mausoleum 378
    5. Inventory of Yount Estate/ Mildred Yount Wealthiest Girl in Texas 381
    6. Mildred Yount's Wedding, June 27, 1938:
       a. *The Tulsa Tribune,* Oklahoma 384
       b. *The Beaumont Journal,* Texas 385
       c. *The Lexington Herald,* Kentucky 387
    7. The Spindletop Dispersal 389

PASSIONS AND PREJUDICE

8. Spindletop Hall and Farm Sold to the University of Kentucky 393
9. Pansy Yount's Obituary 396
10. Pansy Yount's Bequests 398
11. Auction of Pansy Yount's Possessions 399

CORE DOCUMENTS

A. COURT RECORDS

1. W.C. Grant vs. Pansy M. Grant for Divorce

W.C. GRANT	IN THE CRIMINAL
	DISTRICT COURT
VS.	
	OF JEFFERSON
PANSY M. GRANT	COUNTY, TEXAS

TO THE HONORABLE JUDGE OF SAID COURT:

Now comes W.C. Grant, hereinafter called plaintiff, who resides in Jefferson County, Texas, complaining of and against Pansy M. Grant, hereinafter called defendant, who resides in Jefferson County, Texas, and praying for a divorce says and shows to the Court the following:

1.

Plaintiff is and has been for more than twelve months immediately prior to the filing and exhibiting of this petition an actual bona fide inhabitant of the State of Texas, and has resided in Jefferson County, Texas, where this suit is filed, for a period of six months next preceding the filing hereof.

2.

On or about September 27, 1949, plaintiff and defendant were duly and legally married and continued to live together as husband and wife until on or about March 1, 1959.

3.

For a considerable period of time prior to the filing of this petition the defendant, disregarding the solemnities of her marriage vows and her obligation to treat the plaintiff with kindness and attention, and within the past year, commenced a course of unkind, harsh, and cruel conduct towards plaintiff which has continued with very slight intermission until recently. On diverse occasions defendant was guilty of excesses, cruel treatment and outrages toward plaintiff of such nature as to render their further living together insupportable. Plaintiff says that the defendant has been and is now making serious threats of bodily harm against the plaintiff and that upon one occasion the defendant has armed herself with a .380 automatic pistol and that upon one occasion the plaintiff has been forced to disarm the defendant. Plaintiff further says that the

CORE DOCUMENTS

defendant has entered his office and there on more than one occasion has destroyed personal property belonging to him. Plaintiff says that the defendant is possessed of a wild and ungovernable temper and that she should be temporarily restrained from molesting or harassing him during the pendency of this suit.

4.

Plaintiff says that he and the defendant are the joint and mutual owners of real and personal property valued on October 31, 1958, at FOUR MILLION, THREE HUNDRED EIGHTY-FOUR THOUSAND, ONE HUNDRED NINETY-FIVE and 92/100 ($4,384,195.92) DOLLARS. That said property is located in the State of Texas and in several other states. That a portion of said property is separate property of the plaintiff, some of it is separate property of the defendant and the large portion of said estate is community property, and in order that the Court may properly divide this estate, plaintiff says that it is necessary that an independent auditor be appointed by the Court in accordance with Rule 172 of the Texas Rules of Civil Procedure.

5.

Plaintiff says that the complexity of the estate involved in this divorce proceeding requires the appointment of a master in chancery who shall perform all the duties required of him by the Court, and shall be under orders of the Court and have such power as the master of chancery be given in a court of equity and plaintiff prays that said master in chancery be given full powers to investigate and report all the facts concerning the real and personal property owned or held by either plaintiff or defendant and that he be granted all the powers available to a master in chancery as provided by Rule 171 of the Texas Rules of Civil Procedure. Plaintiff says that the defendant is threatening to dispose of all or a part of the community estate belonging to the parties and that she should be immediately restrained from secreting, selling, mortgaging, transferring, giving away, or in any manner disposing of any property whether real or personal now held by her either in her name or in the joint name of the parties hereto. Further in this connection, plaintiff says that the parties have on deposit substantial funds in The American National Bank of Beaumont, Texas, and in The First National Bank of Beaumont, Beaumont, Texas, and that both said banking institutions should be immediately restrained from disposing of or allowing the withdrawal of any funds or personal property standing

CORE DOCUMENTS

in the name of the defendant or any account in which her name appears.

WHEREFORE, plaintiff prays that the defendant be cited to appear and answer herein and that upon final hearing hereof that plaintiff be granted a divorce and that he be awarded his full share of the community estate and that title to his separate estate be quieted in him and plaintiff further prays that the defendant be immediately restrained from molesting or harassing him during the pendency of this suit and that the defendant be immediately restrained from withdrawing funds, secreting, selling, mortgaging, transferring, giving away or in any manner disposing of any property whether real or personal now held by her in her name or in the joint name of the parties hereto during the pendency of this suit and that the First National Bank of Beaumont, Beaumont, Texas, and The American National Bank of Beaumont, Beaumont, Texas, be served with a true copy of said order so restraining the defendant, and plaintiff further prays that the defendant be cited to appear and show cause, if any she can, (a) Why said temporary restraining order should not be held in force during the pendency of this suit; (b) Why an auditor should not be appointed; (c) Why as master in chancery should no be appointed, and plaintiff further prays that he be awarded such other and further relief to which he may be entitled either in law or inequity.

W.C. Grant

W.C. Grant, Plaintiff

ATTORNEYS FOR PLAINTIFF

THE STATE OF TEXAS
COUNTY OF JEFFERSON

BEFORE ME, the undersigned authority, on this day personally appeared W.C. Grant, who having been duly sworn, says that the facts contained in the foregoing pleading are true and correct.

WITNESS MY HAND AND SEAL OF OFFICE this <u>2nd</u> day of March, 1959.

NOTARY PUBLIC in and for Jefferson County, Texas

CORE DOCUMENTS

ORDER OF THE COURT

BE IT REMEMBERED that the above pleading and application for temporary restraining orders was presented to me in open court this 3rd day of March, 1959, at 10:05 a.m., and the Court having considered said pleadings and being of the opinion that unless immediate temporary relief is granted to the plaintiff that irreparable injury will result, the defendant is therefore and hereby immediately restrained (a) From molesting or harassing the plaintiff; (b) From secreting, selling, mortgaging, transferring, giving away or in any manner disposing of any property whether real or personal standing in her name, including funds on deposit in The American National Bank of Beaumont, Beaumont, Texas and The First National Bank of Beaumont, Beaumont, Texas, and the clerk of this court is hereby directed to issue notice to the defendant to be and appear before this Court at 10:00 a.m. on March 12, 1959, to show cause, if any she can, (a) Why said temporary restraining should not be held in force during the pendency of this suit; (b) Why an auditor should not be appointed as prayed for; (c) Why a master in chancery should not be appointed as prayed for. The clerk is further directed to serve a true copy of this order on The American National Bank of Beaumont,

PASSIONS AND PREJUDICE

Beaumont, Texas and The First National Bank of Beaumont, Beaumont, Texas, in order that each of them may be put on notice of the terms and provisions of this order. This being a suit for divorce the Court in its discretion under the terms of 693a TRCP holds in its discretion that no bond be required of the plaintiff in connection with the ancillary relief sought herein.

Judge Presiding

CORE DOCUMENTS

2. **Pansy Grant vs. Cape Grant: Divorce Countersuit/Decree:**

Pansy M. GRANT	IN THE CRIMINAL
	DISTRICT COURT
vs.	
	OF JEFFERSON
W.C. GRANT	COUNTY, TEXAS

DEFENDANT'S ORIGINAL ANSWER

TO THE HONORABLE JUDGE OF SAID COURT:

COMES NOW the defendant, Pansy M. Grant, in the above entitled and numbered cause and files this, her original answer to plaintiff's original petition on file herein, and with respect alleges:

1.

The defendant denies each and every, all and singular, the material allegations in the plaintiff's original petition contained and demands strict proof thereof.

PASSIONS AND PREJUDICE

WHEREFORE, PREMISES CONSIDERED, defendant prays that the plaintiff take nothing by reason of this suit herein and that the defendant go hence without delay and recover her costs.

Respectfully submitted,

ORGAIN, BELL & TUCKER
BALDWIN & GOODWIN
STRONG, MOORE, PIPKIN, STRONG & NELSON

By

[signature: Moore]

Attorneys for Defendant,
Pansy M. Grant

19 McFaddin Bldg.
Beaumont, Texas

COMES NOW, Pansy M. Grant, Defendant in the above entitled and numbered cause, becoming Cross-Plaintiff herein, complaining of W.C. Grant, Plaintiff in

CORE DOCUMENTS

the above entitled and numbered cause, hereinafter called Cross-Defendant, and subject to Defendant's foregoing original answer in said cause and without in any manner waiving same, files this, her original cross-action and would respectfully represent unto the Court the following:

1.

Cross-Plaintiff is now and has been an actual bona fide inhabitant and resident of the State of Texas for more than one (1) year immediately prior to the filing of this suit, and has continuously resided in Jefferson County, Texas for six (6) months next preceding the filing hereof. Cross-Defendant is a resident of Jefferson County, Texas, presently residing at the Ridgewood Motel, Beaumont, Texas.

2.

Cross-Plaintiff and Cross-Defendant were duly and legally married on or about September 27, 1949, and they continued to live and reside together as husband and wife until on or about March 1, 1959, when Cross-Defendant left Cross-Plaintiff and thereafter on March 3, 1959, filed suit for divorce against her. They have not lived and resided together as husband and wife since said separation date, but have lived separate and apart, and the said marriage still exists.

3.

Cross-Plaintiff alleges that during the time she and Cross-Defendant lived and resided together as husband and wife, she was at all times kind, dutiful and affectionate towards the Cross-Defendant, and that she did everything within her power to make their home life happy and pleasant, and to make their marriage a success. Cross-Plaintiff further alleges that prior to their separation Cross-Defendant began a course of harsh and cruel treatment and conduct towards the Cross-Plaintiff so as to render their further living together as husband and wife insupportable.

4.

Cross-Plaintiff alleges that there were no children born or adopted of this marriage.

5.

Cross-Plaintiff further alleges that prior to her marriage to Cross-Defendant her name was Pansy M. Yount; that long prior to her marriage to Cross-Defendant she was the wife of M. F. Yount; that she was the wife of M.F. Yount continuously from the time of her marriage to him until his death in 1933; that she did not at any time marry anyone else prior to her marriage to Cross-Defendant; that she now petitions the Court that the name of Yount be restored to her

CORE DOCUMENTS

so that after this divorce is final she will be legally known as and called "PANSY M. YOUNT."

6.

Cross-Plaintiff would further show unto the Court that at the time of her marriage to Cross-Defendant she owned in her own right, as her separate property and estate, a large amount of real and personal property; that all of said property had been acquired during her marriage with M.F. Yount who was her husband prior to her marriage to Cross-Defendant; that her prior husband, M.F. Yount, became deceased in 1933 and following his death there was a partition between Cross-Plaintiff and the only child of her marriage with M.F. Yount; that the property owned by her as her separate estate at the time of her marriage to Cross-Defendant was the property partitioned to her following the death of her said prior husband, and certain additional property acquired by her during the period from the death of her prior husband and her marriage to Cross-Defendant; that she was the sole owner and holder of all the title to all such property at the time of her marriage to Cross-Defendant; that such property continued to be and is now her separate property and estate.

PASSIONS AND PREJUDICE

That during the entire period of time she has been married to Cross-Defendant, to wit, from September 27, 1949, to date, the entire income from the aforesaid property from all sources, and more, has been expended by said Cross-Defendant and Cross-Plaintiff; that if property has been acquired by this Cross-Plaintiff since her marriage to Cross-Defendant said property was acquired with funds from the separate property and estate of this Cross-Plaintiff, and therefore constitutes the separate property of Cross-Plaintiff.

That this Cross-Plaintiff should have judgment against Cross-Defendant vesting in her the full title and possession of all property owned by her at the time of her marriage to Cross-Defendant, and all property thereafter acquired by her with funds derived from her separate property and estate.

This Cross-Plaintiff should have judgment against Cross-Defendant vesting in her the full title and possession of all property owned by her at the time of her marriage to Cross-Defendant, and all property thereafter acquired by her with funds derived from her separate property and estate.

This Cross-Plaintiff further alleges that during her marriage with Cross-Defendant a great amount of property has been

CORE DOCUMENTS

accumulated by the Cross-Defendant in his name, but that said property is in fact the separate property of this Cross-Plaintiff; that if Cross-Plaintiff be mistaken in such allegation that such property now held by Cross-Defendant in his name is the community property of this Cross-Plaintiff and Cross-Defendant; that this Cross-Plaintiff should have judgment against said Cross-Defendant fully vesting in her the title and possession of all such property now held by the Cross defendant in his name; or in the alternative should have judgment against Cross-Defendant to the extent of her interest in any such property held by the Cross-Defendant in his name.

WHEREFORE, PREMISES CONSIDERED, Cross-Plaintiff prays that the Cross-Defendant, W.C. Grant, be cited to appear and answer herein, that upon a final trial and hearing hereof the Cross-Plaintiff, Pansy M. Grant, have judgment (a) for a divorce dissolving said marriage on her cross-action; (b) ordering that the name of "YOUNT" be restored to Cross-Plaintiff so that she will be legally known as and called PANSY M. YOUNT; (c) vesting in her title and possession of all property, real and personal, owned by her at the time of her marriage to Cross-Defendant and all property, real and personal, thereafter acquired by her; (d) vesting title in her as her separate property and estate of all property standing in the name of Cross-Defendant, or in the alternative, vesting title to an undivided one-half (1/2) interest in said property standing in the name of Cross-Defendant; (e) vesting in Cross-Plaintiff title to all furniture, furnishings and household goods, silver, china, paintings, bric-a-brac, draperies, rugs, jewelry, cash in bank, stocks and bonds, and all other personal property of every kind claimed by this Cross-Plaintiff as her separate property and estate; (f) ordering Cross-Defendant to pay all court costs in this behalf expended; and Cross-Plaintiff prays for such other and further relief, general and special, legal and equitable, to which Cross-Plaintiff may show herself justly entitled.

Respectfully submitted

ORGAIN, BELL & TUCKER
BALDWIN & GOODWIN
STRONG, MOORE, PIPKIN,
STRONG & NELSON

BY *A.D. Moore*
Attorneys
for Cross-Plaintiff,
Pansy M. Grant
319 McFaddin Bldg.
Beaumont, Texas

CORE DOCUMENTS

A copy of the above and foregoing Defendant's Original Answer and Cross-Plaintiff's Cross Action has been forwarded to Mr. W.G. Walley, Jr., Bowie Building, Beaumont, Texas, and Mr. Everette Lord, American Nation Bank Building, Beaumont, Texas, Attorney's for Plaintiff and Cross-Defendant herein.

Entered December 17, 1959

W.C. GRANT	IN THE DISTRICT COURT
vs.	OF JEFFERSON COUNTY, TEXAS 136TH JUDICIAL
PANSY M. GRANT	DISTRICT

DECREE

On the 26th day of October, 1959, came on to be heard the above entitled and numbered cause wherein W.C. Grant is Plaintiff and Cross-Defendant, and Pansy M. Grant is Defendant and Cross-Plaintiff; and

PASSIONS AND PREJUDICE

came the said parties, W.C. Grant, Plaintiff and Cross-Defendant, and Pansy M. Grant, Defendant and Cross-Plaintiff and announced ready for trial; and came a jury of twelve good and lawful men, who, being duly impaneled and sworn and having heard the pleading, the evidence and the argument of counsel, on their oaths and in response to the special issues, definitions and explanatory instructions submitted to them by the Court, on the 17th day of November, 1959, make the following respective findings:

SPECIAL ISSUE NO. 1

Do you find from a preponderance of the evidence that the acts of the defendant, Pansy M. Grant, toward the Plaintiff, W.C. Grant, if any, constitute such excesses, cruel treatment or outrages of such a nature as to render their further living together as husband and wife insupportable.?

You will answer "We do" or "We do not."

Answer: "We do not."

SPECIAL ISSUE NO. 2

Do you find from a preponderance of the evidence that the acts or conduct, if any, of

CORE DOCUMENTS

W.C. Grant toward Pansy M. Grant constitutes such excesses, cruel treatment or outrages of such a nature as to render their further living together as husband and wife insupportable as that term is defined in this charge?

You will answer "We do" or "We do not."

Answer: "We do."

And the court having adopted the findings of the jury as to such Special Issues Nos. 1 and 2, it is ORDERED, ADJUDGED, and DECREED by the Court that the prayer and request of the Plaintiff, W.C. Grant, for a divorce from the Defendant, Pansy M. Grant, be and it is hereby denied; and it is further ORDERED, ADJUDGED, and DECREED by the Court that the Cross-Action of the Defendant, Pansy M. Grant, for a divorce from the Cross-Defendant, W.C. Grant, be granted and that upon the prayer and request of Cross-Plaintiff, Pansy M. Grant, the bonds of matrimony heretofore existing between Cross-Defendant, W.C. Grant, and Cross-Plaintiff, Pansy M. Grant, be and they are hereby dissolved, and the Cross-Plaintiff, Pansy M. Grant, is granted a divorce from the Cross-Defendant, W.C. Grant.

PASSIONS AND PREJUDICE

And it further appearing to the Court that the Cross-Plaintiff, Pansy M. Grant, has requested that her former name of "Pansy Merritt Yount" be restored; and the Court having found that such should be done:

It is therefore ORDERED, ADJUDGED, and DECREED by the Court that the Cross-Plaintiff, Pansy M. Grant, be and she is hereby restored her former last name of "YOUNT' so that hereafter she will be legally known as and called by the name of "Pansy Merritt Yount," although in this Decree she is hereinafter referred to as "Pansy M. Grant," such as being the name under which she appears of record in the pleadings herein.

And it further appearing to the Court, and the Court having found, under the undisputed evidence, that those certain properties hereinafter described, each and all, constitute and are the separate property and estate of the Defendant and Cross-Plaintiff, Pansy M. Grant, to wit:
SIGNED, RENDERED, ENTERED AND FILED this 17th day of December, A.D. 1959.

HAROLD R. CLAYTON,
JUDGE PRESIDING

CORE DOCUMENTS

B. NEWSPAPER CLIPPINGS

Articles on Pansy Yount, Miles Frank Yount, Mildred Yount, Spindletop, Spindletop Hall, Cape Grant, and Spindletop's Saddlebred Horses.

PASSIONS AND PREJUDICE

1. Stradivari Violins

THE BEAUMONT ENTERPRISE, 1932

YOUNT BUYS HIS DAUGHTER MILLION DOLLAR COLLECTION OF WORLD'S FINEST VIOLINS

Mildred Yount, the 11-year old daughter of Mr. and Mrs. M.F. Yount of Beaumont, Texas, and Manitou, Colorado, has become the owner of the foremost collection of old violins in America. Her father has bought them for her, it is reported at a cost of at least $1,000,000. The little girl is herself studying the violin and has shown remarkable talent. In the collection are listed the following violins, given in the order of maker, name, and date:

Joseph Guarnerius, "Del Jesu," 1741; Antonio Stradivari, "Spanish," 1689; Stradivari "Swan," 1737; Stradivari, "Reynier," 1684; Stradivari, "Piatti," 1717; Dominicus Montagnana, 1735; J.B. Guardagnini, 1782; Andre Guarnerius and Felius Andre, 1715.

On March 7 the Texas oil magnate and Mrs. Yount and their daughter Mildred, presented Paul Kochanski, the famous virtuoso, in a recital at their Beaumont home and played the numbers of the program, each on a different violin. It was violin history.

CORE DOCUMENTS

There are few great artists in the world who have as good violins, it is said, as any one of these famous violins that this 11-year-old girl owns. And they are not to her only a collection of relics. A Student of the violin, their marvelous tones, which no modern hands, however cunning, can duplicate, are hers to command. The priceless fiddles, some of which have probably delighted kings and which have moved human feelings throughout hundreds of years, such as only a violin maker's masterpiece in the hands of a master player can, repose each in its beautiful case for her to play at will.

Mr. Yount is said to have a passion for music and it is reported to be Mrs. Yount's ambition to see their daughter become an artist. Mildred has shown rare talent and is said to practice assiduously. She has a wealth of violins probably never equaled by any child or adult violinist in the world, the nobility and royalty of Europe excepted.

PASSIONS AND PREJUDICE

2. Miles Frank Yount Honored by Rotary in Beaumont:

THE BEAUMONT ENTERPRISE, JUNE 14, 1933

YOUNT IS HONORED AS DISTINGUISHED CITIZEN BY ROTARY

Beaumont Rotarians paid high tribute to one of the city's outstanding citizens, Miles Frank Yount, at their luncheon-meeting Wednesday noon at Hotel Beaumont, with J.S. Edwards praising him as a public benefactor in the chief address of the noon hour.

Edwards said that Beaumont and the Rotarians in particular held Mr. Yount in high esteem and admiration. "Honor to whom honor is due," Edwards said at the offset and proceeded to praise the honor guest for his good citizenship, which called for civic pride, liberality to the poor and high business ethics. Mr. Yount, the speaker said, had rallied to the cause of his friends and Beaumont more than once, and had given his time and means for up building the city.

He praised Mr. Yount as a good neighbor, a friend of everyone, and a man who has been of untold benefit to the city. Following his brief remarks, Edwards

CORE DOCUMENTS

presented Mr. Yount with a scroll from the club, setting forth the virtues of the Rotary Club's outstanding citizen of Beaumont.

Mrs. Yount and daughter, Mildred, were guests of the club, Mrs. Yount being presented with a large basket of roses.

President C.A. Easley presented Mr. Yount, his wife, and daughter to the club.

The full text of the scroll presented to Mr. Yount follows:

TESTIMONIAL

The phenomenal career of Miles Frank Yount offers a striking illustration of the unlimited opportunities which our blessed country with its vast resources affords to the man who has the energy and ability to make the right use of them.

Miles Frank Yount is an inspiring example of the self-made man who has risen from the ranks to a commanding position, not by the favor of fickle fortune, but by dint of hard work, close application and untiring effort. He owes his signal success not to blind chance and lucky strikes, but expert knowledge, gained in the school of experience. If the chosen field of his labors, the oil industry, has enriched him, he, in turn, has enriched the mechanics of that industry by hundreds of invaluable devices.

PASSIONS AND PREJUDICE

However, this testimonial is not dedicated to Miles Frank Yount the oil operator, but to Miles Frank Yount the public-spirited citizen, the community building, and, above all, to the generous-hearted man who uses his wealth not for self-aggrandizement, but the welfare of his fellowmen.

Miles Frank Yount loves Beaumont and has unbounded faith in its future. He has been untiring in promoting the interests of our city, and in furthering its growth and prosperity. He has given generously, though unostentatiously, to every worthy cause, fostered by the institutions of our city. Unlike a beacon which illumines distant points, but has its base shrouded in darkness, Miles Frank Yount is large-hearted to his employees and is fully mindful of the large share, which their loyal co-operative effort has in the success of his enterprises.

The Rotary Club, founded as it is on the principles of unselfish service, feels that the distinguished career of Miles Frank Yount should receive public recognition, and be held up as a model for others to imitate and emulate.

With this end in view, The Rotary Club of Beaumont confers upon Miles Frank Yount the title of

DISTINGUISHED CITIZEN

CORE DOCUMENTS

and accompanies this honor with the prayer that Providence may grant him to enjoy a long life and enable him to continue, in an ever-increasing measure, his fruitful civic and humanitarian service.

PASSIONS AND PREJUDICE

3. Funeral Miles Frank Yount; Horse Show Goes On

Beaumont Enterprise
November 15, 1933

YOUNT HORSES TO APPEAR IN SHOW

SPINDLETOP STABLE ENTRIES WILL COMPETE AS OWNER HAD PLANNED

With their red and blue colors proudly flying, six fine horses from the Spindletop Stable of M.F. Yount will be on exhibit at the Kansas City Royal Horse Show Saturday, just as their owner had planned for them to be.

For a number of months prior to his sudden death here Monday night, Mr. Yount had probably experienced more enjoyment from the Yount horses than from any other of his numerous avocations or hobbies, and had looked forward to attending the Kansas City show with keen anticipation.

His funeral will be held at the very hour he had expected to be in route to Kansas City, but members of the family believe it would be in accordance with his wishes for his exhibit to go through as he had originally planned.

CORE DOCUMENTS

Cape Grant, in charge of Spindletop Stables, has been in Kansas City with the horses for several days, and will supervise their showing. Beau Peavine, Night Alarm, Chief of Spindletop, Lady Virginia, Blue Bonnet McDonald, and Whirlwind McDonald--all will be shown.

PASSIONS AND PREJUDICE

4. Yount Mausoleum

Beaumont Enterprise

FRANK YOUNT MAUSOLEUM MAY BE ERECTED OF STONE TAKEN FROM OWN QUARRIES

Colorado Greenstone Being Considered; Services Are This Afternoon At 3 o'clock in Home

The body of Miles Frank Yount, 53, Beaumont oil man who died Monday night, will be carried to the grave this afternoon by 12 oil field employees of his company. Already plans are being made for the construction of a mausoleum in Magnolia cemetery to hold the body permanently. If it is decided that the Colorado green stone from the Yount quarry is suitable for construction of the mausoleum, this material will be used.

Bronze is considered as an alternative.

Mr. Yount will be buried this afternoon in the Merritt lot on the south side of the Magnolia Cemetery, the body to be removed later to the mausoleum.

Yount's Mother Arrives

Mrs. Hattie Minerva Yount, 84, mother of Mr. Yount, arrived here yesterday morning *from her home in Monticello, Arkansas, birthplace of her son. She was accompanied here by a grandson, Harrell Yount.*

Funeral services will be held in the residence, 1376 Calder Avenue, this afternoon at 3 o'clock, and a public address system will be installed so that the crowds that cannot be accommodated inside the home can hear the service from the lawn.

Dr. Hunter Officiates

Dr. T.M. Hunter, pastor of Westminster Presbyterian Church, will officiate at the services. Burial will be under direction of Pipkin and Brulin.

The casket, solid cast bronze, and in a solid copper vault, the two weighing more than a ton, arrived last night at Pipkin and Brulin, having been ordered specially for the Yount funeral.

About 20 men, more than half of whom were employees of the oil company, were called upon to unload the casket and vault from the railway company truck. The casket weighs slightly more than 1300 pounds, and the copper vault about 800 pounds.

Activities of the Yount-Lee Oil Company in oil fields of Texas and Louisiana will be suspended today in honor of the man who brought many of the fields into being and who expanded others, and in Beaumont a number of business houses will be closed.

PASSIONS AND PREJUDICE

All banks of the city, through action of the Clearing House Association, will close at 2:30 o'clock this afternoon, half an hour before the funeral services. Offices at the city hall will close at noon out of respect to the man who had turned large sums of money into the city treasury that city employees might not be without pay checks on pay days.

Other Officials

Mr. Yount came to Beaumont at the age of 17, and after years of manual labor built up the Yount-Lee Oil Company, one of the largest independent oil producing companies in the nation.

Associated with him in the company at the time of his death were T.F. Rothwell, vice president; J.H. Phelan, secretary and treasurer; F.E. Thomas, assistant secretary, and Beeman Strong, general counsel, all of Beaumont, and E.F. Woodward, W.E. Lee, and T.P. Lee, of Houston.

CORE DOCUMENTS

5. Inventory of Yount Estate/Mildred Wealthiest Girl in Texas

INVENTORY OF FATHER'S ESTATE SHOWS THAT 13-YEAR OLD MILDRED YOUNT TO BE WEALTHIEST GIRL IN TEXAS

Mildred Yount, age 13, daughter of the late M.F. Yount, is today the wealthiest girl in Texas.

An inventory of her late father's estate was filed yesterday. The entire estate is to be shared between the young woman and her mother.

Virtually all of the huge fortune is in stocks and real estate. These include blocks of stock in the Texas Corporation, the Wall Street Property Company, and the First National Bank of Beaumont. The largest holdings are in the Yount-Lee Oil Company, of which her father was president.

She has few definite plans for the future. Her education will be thorough. She will attend college, probably the University of Texas, of which her father was a regent, and will travel abroad.

She is considered above the average in her school work, and is being trained in music. She can ride beautifully, and her trainer, Cape

PASSIONS AND PREJUDICE

Grant, says that within a few years she will ride in some of the finest horse shows in America.

Mildred was her father's almost constant companion. When he established the now famous Spindletop Riding Stables, which hold several of the finest saddle horses in America, his daughter took a keen interest in them.

Since his death her mother has purchased for her the remarkable horse Roxie Highland, held to be the finest three-gaited saddle horse in the world.

Cranks have recently given Mrs. Yount much annoyance. Several letters, from various parts of the country, have proposed marriage with her 13-year old girl, as well as with Mrs. Yount.

Mildred, a little freckled from the sunshine in which she plays, is being given her education in her home at 1376 Calder Avenue by private tutors.

This is by no means because the young heiress or her mother do not advocate public schools, but rather on account of threats which have been made against her. From the time she was old enough to enter school, Mildred attended public school.

She went to Millard school for her elementary education, but when time came for her to enter junior high school her father and mother decided to keep her at home. This the little girl did not like, but threats were being made, more than ever were being made public. The Lindbergh baby had been kidnapped. There were many notes and telephone calls to the Yount residence. These were communicated to the police.

Carefully Guarded

Mildred Yount was guarded carefully. One bodyguard, a husky man, remained with her constantly. Her father gave her a little automobile of her own, but, though she was large enough to drive it occasionally, the bodyguard had to go with her. Her parents provided police dogs for their home and an intricate burglar alarm system.

"It doesn't seem as though I'm like other little girls," she said. "I can't even play without being watched, and now I can't even go to school with my playmates."

As a consequence, the Yount home has a constant stream of youthful visitors. But Mildred would rather play on the side walk than in her magnificently appointed rooms, or anywhere in her gorgeous home. To her, life has become circumspect and a little drab.

PASSIONS AND PREJUDICE

6. Mildred's Wedding (Newspaper stories)

THE TULSA TRIBUNE, MONDAY, JUNE 27, 1938

MILDRED YOUNT, SOUTHWEST'S RICHEST OIL HEIRESS MARRIES YOUNG TULSA ATTORNEY IN LEXINGTON, KY.

The southwest's richest oil heiress, 18-year old Mildred Frank Yount of Beaumont, Texas, and Spindletop Hall near Lexington, was married in Lexington today to 27-year old Edward Daniel Manion, a tall, dark, and slender young Tulsa attorney.

The marriage unites the heiress to what has been estimated as a $50,000,000 fortune and the son Mr. and Mrs. John A. Manion, 2131 E, Twenty-seventh street, Tulsa. Mr. Manion is vice president and general manager of the Sinclair Refining Co., and pipeline department.

CORE DOCUMENTS

THE BEAUMONT JOURNAL, JUNE 27, 1938

MILDRED YOUNT, OIL HEIRESS, WILL WED THIS MORNING TO TULSA, OKLA., ATTORNEY

Miss Mildred Yount, daughter of Mrs. Pansy Yount of Lexington, Kentucky, and the late Miles Frank Yount, Beaumont oil man will be married at 10 o'clock this morning in the Catholic church of Lexington to Ed Manion of Tulsa, Oklahoma.

After a wedding trip to Europe, they will return wither direct to Beaumont or by way of Manitou, Colorado, where the Yount summer home, "Rockledge," is located. They will arrive in Beaumont about October 1 to make their home and Mr. Manion, a law graduate of Tulsa University, will be associated in legal practice here with R. E. Masterson, personal attorney of the Yount family.

Met in Colorado

It was in Manitou last summer that the romance began between Miss Yount, one of the richest young women in America, and the Oklahoman. Miss Yount completed her studies in the Hockaday school, Dallas, during the last year and was graduated a month ago, along with her long-time, best friend, Dorothy Hilliard, who attended with her and will be her maid of honor.

PASSIONS AND PREJUDICE

Every precaution had been taken to keep the approaching wedding a secret until the last minute, since Mrs. Yount hoped to avoid public attention being centered on the event.

A simple ceremony is planned for this morning with only members of the immediate families of the principals and close friends in attendance.

Mr. Masterson to Serve

The bride is to be given in marriage by Mr. Masterson, who more than being the family's attorney, has for years been one of the closest friends of the Younts. As soon as the arrangements for today's wedding had been decided upon, Miss Yount herself requested Mr. Masterson to participate in the service.

Four priests will officiate in the wedding rites. Two are fathers of the parish of Lexington, one is from the parish to which Mr. Manion belongs in Tulsa, and the fourth will be Msgr. E.A. Kelly of St. Anthony's parish, Beaumont, with which the young couple will become affiliated.

CORE DOCUMENTS

THE LEXINGTON HERALD, JUNE 27, 1938

TULSA, OKLA., ATTORNEY WED TO HEIRESS OF SPINDLETOP OIL MILLIONS AT LEXINGTON

The marriage of Miss Mildred Frank Yount, daughter of Mrs. Miles Frank Yount of Spindletop Farm, Lexington, and El Ocaso, Beaumont, Texas, and the late Mr. Yount, to Mr. Edward Daniel Manion of Tulsa, Okla., son of Mr. and Mrs. J. R. Manion of Tulsa, was impressively solemnized at 11:00 o'clock this morning in St. Paul's Catholic Church, the Reverend Joseph E. McKenna reading the service before the immediate families and a few friends.

The church, of which Mrs. Yount was a member, always sitting in the third row from the front on the right hand side on Sunday mornings, was elaborately decorated with white lilies, Southern smilax, Woodwardia ferns and candles. Along the communion rail a hedge of lilies was arranged. The alter was banked with lilies, interspersed with wedding candles, and on either side were clusters of Woodwardia ferns and lily trees.

Placed around the columns in the church were clusters of ferns; clusters of lilies marked the end of each pew, the entire length of the middle aisle, and white satin streamers divided the rows of seats throughout the

PASSIONS AND PREJUDICE

church. The choir loft and vestibule were decorated with ferns and smilax, while large vases of lilies were placed just inside the vestibule to welcome the lovely bride.

A program of wedding music played as the guests assembled and during the ceremony included:

Processional..Verne
On This Day...Lambilotti
Ave Maria..Rosewig
Oh Lord I Am Not Worthy.....Traditional
Magnificat......................................Traditional
Recessional...Wedding March........Valenti

The bride, escorted to the altar by Judge R.E. Masterson, of Beaumont, Texas, wore a gown especially designed for her wedding by Madame Coonti of Paris and New York. Fashioned of white silk net over white slipper satin, its simple bodice, fastened with tiny satin buttons had a high neckline finished with a round collar of rosepoint and needlepoint lace made in Belgium, was adjusted to a halo of the same lace, lined with tulle and trimmed with little sprays of orange blossoms. Her only ornament was a diamond necklace, the gift of her late father.

CORE DOCUMENTS

7. The Spindletop Dispersal Sale

Lexington Herald-Leader

MISS DIXIE REBEL TOPS DISPERSAL AT SPINDLETOP

A bid of $10,000 for Miss Dixie Rebel, four-year-old fine harness mare, topped the Spindletop Dispersal Sale of saddle and show horses yesterday as 30 head sold for a total of $52,400, an average of 1, 747.

Robert Baskowitz of St. Louis, Mo., purchased the chestnut daughter of Beau Peavine-Abie's Irish Rose by American Born after a brisk bidding duel.

Second high price was the $8,000 while Argyle Stables paid for Spindletop's famed coach and four. The four horses included in this offering were Waltz Time, Dinner Dance, Curtain Call, and Stage Hand.

Abie's Royal Irishman, six-year-old five-gaited stallion by Beau Peavine-Abie's Irish Rose, went to C. Koenig of Philadelphia for $5,000.

Two horses brought $4,500 each, both five-gaited stallions. Lexington Leader, a six-year-old by Bourbon Genius-Belle Le Rose by American Born was purchased by Robert C.

389

PASSIONS AND PREJUDICE

Beatty of Washington, Pa., while Chief of Texas, a three-year-old by Beau Yount-Abie's Irish Rose by American Born went to Louis Greenspon of St. Louis, Mo.

The sale was conducted by Tattersalls, Inc., which also sold 24 head at the Tattersalls Sales Barn yesterday morning. George Swinebroad was the auctioneer.

Sales will resume at Tattersalls this morning.

Summary of the Spindletop Sale:

Rose of Spindletop, George Gwinn, Danville, Ky., $300.

The Virginia Princess, Eli Long, Delaware, Ohio, $400.

Little Rose Delight, Elizabeth Kittell, Nashville, Tenn., $1,000.

Spindletop Princess, Lee B. Thomas, Anchorage, Ky., $300.

Lexington Leader, Robert C. Beatty, Washington, Pa., $4,500.

Spindletop Ike, R.L. Mansell, Medina, Ohio, $150.

Lady Lafitte, George Gwinn, Danville, Ky., $650.

Miss Dixie Rebel, Robert Baskowitz, St. Louis, Mo., $10,000.

Floy Watkins, Ike Lanier, Danville, Ky., $450.

Sissy Peavine, F.H. Eddy, Lexington, $150.

CORE DOCUMENTS

Mildred Manion, Lewis C. Terney, Charleston, W. Va., $1,450.

Sophie Van Cleve, R.C. Tway, Louisville, Ky., $800.

Peggy Odom, Howard King, Lexington, $300.

Silver Masterpiece, Thomas Murphy, Danville, Ky., $1,400.

Nancy's Beau, Louis Greenspon, St. Louis, Mo., $2,700.

Abie's Royal Irishman, G. Koenig, Philadelphia, Pa., $5,000.

Spindletop Bourbon, James E. Krueger, Cleveland, Ohio, $150.

Ducky West, H.S. Whittenburg, Louisville, Ky., $100.

Chief of Texas, Louis Greenspon, St. Louis, Mo., $4,500.

Marie Bosace, Robert E. Boettcher, Springfield, Ohio, $800.

Grassland's, T.D. Adams, Indianapolis, Ind., $1,600.

Susie Kettman, George Gwinn, Danville, Ky., $400.

Abie's Baby, Ike Lanier, Danville, Ky., $1,000.

Kathryn Manion, T.D. Adams, Indianapolis, Ind., $1,000.

Roxie Lafitte, J. Howard, Lexington, $850.

Kalarama Kathleen, W.W. Evans, Jeffersontown, Ky., $1,150.

PASSIONS AND PREJUDICE

Inky Imp, H.M. Boock, Logan, Ohio, $1,600.

Miss Monroe, Ed Malerick, Lincoln, Ill., $1,400.

Coach and Four, Argyle Stables, $8,000.

CORE DOCUMENTS

8. Spindletop Hall and Farm Sold to the University of Kentucky

LOUISVILLE COURIER-JOURNAL
February 21, 1959

SHOW-PLACE FARM BOUGHT FOR U. OF K.

BARGAIN $850,000 PAID FOR 1,066-ACRE SPINDLETOP; MANSION ALONE WORTH MORE

Lexington, Ky., Feb. 20.--Spindletop Farm, a 1,066-acre show place near Lexington with a three-story mansion valued alone at more, was bought Friday by the Kentucky Research Foundation for $850,000 as an investment.

The foundation administers all of the University of Kentucky's research activities.

An agreement to buy the well-known farm, on Iron Works Pike about five miles from downtown Lexington, was reached after a meeting of the foundation's board of directors and the executive committee of the U.K. board of trustees.

PASSIONS AND PREJUDICE

STATE FUNDS EXPECTED

U.K. Vice-President Leo M. Chamberlain, president of the foundation, said, "No state funds are involved in the transaction at this time." He issued a formal statement which said:

"It is anticipated that $150,000 will be made available by the State on July 1, 1959, to be applied on the purchase price of $850,000. The remaining price of $700,000 will be an obligation of the Kentucky Research Foundation."

Chamberlain's statement described the purchase as a "step which will be of immeasurable value to the University of Kentucky. There are several factors related to this transaction which deserve comment.

AN INDEPENDENT CORPORATION

First, the Kentucky Research Foundation is an independent corporation and as such is legally separated from the University and other universities, sometimes through multimillion-dollar investments made by the foundation for the benefit of the institution.

'HAS EXCELLENT LAND'

"Second, the University of Kentucky already has a large amount of excellent land available for research and experimentation. However, during the time which the Kentucky Research Foundation may hold this land, the College of Agriculture will supervise farming activities on Spindletop Farm."

CORE DOCUMENTS

The Kentucky Agricultural Experiment Station "has been asked to stock and operate the farm," Chamberlain said. "As a result of this arrangement, the University of Kentucky, including the experiment station, will benefit from the acquisition of this property at a price currently far below market value.

U.K. 'EXTREMELY FORTUNATE'

"As has been indicated in newspaper reports, the transaction represents a sizable 'gift' to the university. The university considers itself extremely fortunate to be the beneficiary of such an investment as this one made possible through the Kentucky Research Foundation and the Commonwealth of Kentucky."

Governor Chandler told a luncheon club audience on January 30 that the farm had been offered to U.K. for $850,000, and the university accepted.

Present at the joint meeting of the trustees and the foundation directors were President Frank G. Dickey, Chamberlain, Robert P. Hobson, and Harper Gatton, Louisville; Dr. Ralph Angelucci, J. Stephen Watkins, Frank D. Peterson, Dr. E.N. Fergus, and Prof. Wilburt D. Ham, all of Lexington.

PASSIONS AND PREJUDICE

9. Pansy Yount's Obituary

THE BEAUMONT ENTERPRISE
OCTOBER 15, 1962

MRS. PANSY YOUNT WIDOW OF OILMAN DIES AT RESIDENCE

Mrs. Pansy M. Yount, widow of Miles Frank Yount, discoverer of the second field at Spindletop in the mid-twenties, died at 11:34 a.m. yesterday in her residence following an illness. She was a native of Orange but had spent most of her life in Beaumont.

During years when her husband headed the Yount-Lee Oil Co., the philanthropies of the couple to the city and community were legend. Mr. Yount died of a heart attack in 1933 and the firm was sold to Stanolind Oil Co. for $46,000,000.

Kentucky Farm

Following her husband's death, Mrs. Yount acquired a farm in the blue grass country of Kentucky. With the farm went the priceless collection of early American horse-drawn vehicles.

Failing Health

Due to her failing health, Mrs. Yount had retired from all outside activities.

CORE DOCUMENTS

She is survived by a daughter, Mrs. E.D. Manion of Beaumont; three grandchildren, Mrs. Thomas Haider of Chicago, Miss Mildred Yount Manion and Edward Daniel Manion, Jr., both of Beaumont, and a brother, Travis J. Merritt of Beaumont.

Her body will remain at the Caldwood residence, where a Rosary will be recited at 7:30 p.m. today.

Requiem Mass will be said by Msgr. E.A. Holub, pastor, in St. Anne's Catholic Church at 10 a.m. tomorrow. She was a member of St. Anne's.

Burial will be in Magnolia Cemetery beside her beloved husband. Arrangements are in charge of Broussard Mortuary.

Pallbearers will be Fred B. Wachs, Lexington, Kentucky; Joe A. Fisher, New York City; M.R. Geisendorff, Ed Fitzpatrick, W.S. Yount Jr., Harry M. Heffner, Jr., Sam S. Roberts, and John E. Gray.

PASSIONS AND PREJUDICE

10. Pansy Yount's Bequests

THE HOUSTON POST, OCTOBER 24, 1962

U.T., MASONIC HOME TO BENEFIT IN YOUNT WILL

BEAUMONT--Bequests in the estimated multi-million dollar will of Mrs. Pansy M. Yount of Beaumont have been made to the University of Texas, the Masonic Home for Crippled Children at Lexington, Kentucky, several grandchildren, a "friend," and "one who has served me faithfully."

The will of Mrs. Yount, 75, wife of Miles Frank Yount, discoverer of the second oil field at Spindletop in the mid-20's, was filed for probate in the Jefferson County Court here. She died at her home on October 14.

The University of Texas and the Crippled Children's Home are to benefit from jewelry and art treasures in the estate.

The grandchildren and their bequests include Mildred Yount Manion of Beaumont, a homestead; Edward Daniel Manion Jr., of Beaumont, $200,000; and Kathryn Bernadette Manion Haider of Chicago, $200,000.

Other beneficiaries include Mrs. Viola G. Ransom of Beaumont, a "faithful friend," $50,000, and Ed Fitzpatrick of Beaumont, "who has served me faithfully," $25,000.

CORE DOCUMENTS

11. Auction of Pansy Yount's Possessions

THE HOUSTON CHRONICLE, JANUARY 14, 1964

MRS. YOUNT'S POSSESSIONS CATALOGUED FOR AUCTION

The personal possessions--the grand and the insignificant--belonging to a famous Texas family will go on auction block beginning Sunday.

THE ESTATE OF the late Mrs. Miles Frank Yount of Beaumont, whose husband was one of the founders of the Yount-Lee Oil Company, is being sold by the executors.

Samuel Hart Galleries will auction the 3,000 items in a two-week sale--the most extensive the gallery has ever handled.

WHEN MRS. YOUNT died October 14, 1962, at 75, her will left bequests estimated at $1 Million.

Proceeds of the sale will go to Mrs. Yount's granddaughter, Mildred Yount Manion of Beaumont. Miss Manion is the daughter of Mrs. E.D. Manion of Beaumont.

The collection of furniture and accessories has been gathered in Houston

PASSIONS AND PREJUDICE

from the far corners of this country and includes the contents of six homes.

THE ITEMS COME from four mansions--Spindletop Hall in Lexington, Kentucky; Woodland Estate in Florida; El Ocaso, and Caldwood Drive Estate in Beaumont, a summer home, Rockledge in Manitou Springs, Colorado, and a bay house in La Porte, Louisiana.

The staggering job of taking inventory and cataloguing these possessions has been the task of Mrs. Trent Newton of Beaumont, an old family friend and secretary for the estate. Public exhibition will be held from 1 to 8 P.M. Saturday at 5510 Greenbriar.

"I'm here because I know what goes with what. I spent as much time in the Yount home as I did in my own," Mrs. Newton said, leading a tour of the block-long building where everything was in various stages of uncrating.

Her task has taken two months. She described many of the pieces she recognized, most of which reflect a phase of the lives of the people "who were interested in everything."

There is a melodeon which belonged to Baby Doe Tabor, a romantic and tragic heroine of the gold rush days of Colorado.

ALSO REMINISCENT of the family's life in the West is a collection of about 100 Navajo Indian rugs.

CORE DOCUMENTS

An unusual collection contains 14 or 15 mounted Texas Longhorns. Mrs. Yount had so many, because there were Longhorns at her home in the West and at her Kentucky home, Mrs. Newton explained.

The Yount horses were famous in the show rings throughout the country. The sterling silver trophies they won will be sold.

Also listed are gold-banded plates of Lenox china, each with the picture of a famous Yount horse in the center. Among those are Champion Beau Peavine, Chief of Spindletop, and Roxie Highland, grand champion three-gaited mare.

THE CHANDELIERS have arrived--each denuded of crystal prisms, all of which must be assembled before the sale.

"We have thousands of prisms and it will take hours," Samuel Hart said. Hart said items will probably go for as little as $1 (a set of everyday glasses) or as much as $10,000 (a made-to-order, 20-piece dining room suite of heavily carved tables, side boards, cabinets with marble tops, and twelve chairs).

The most valuable items probably are 200 or so Oriental rugs, a set of six rare bronze Sarouk elephants and a collection of carved jade, Hart said.

"The family had pianos almost everywhere you looked," Mrs. Newton said, and several will be sold.

PASSIONS AND PREJUDICE

A PIANO THAT will not be sold is one which Paderewski played when he visited the Beaumont home while on concert tour.

There are enough linens to stock 50 small homes, Hart said. These include handmade sheets.

"I can't tell you how many handmade quilts we have," Mrs. Newton said.

Family furs will be sold including three boxes of fur-collared coats. The boxes are marked "second floor closet."

ABOUT 100 PIECES of copper serving pieces, camel bells, Catholic religious articles, paintings, mirrors, tables, beds, and a spinning wheel are among the things stacked in the auction room.

"One set of Spode china has 54 cups and saucers, Mrs. Newton said. "When Mrs. Yount bought anything, she always bought a set, at least 12 of everything and an extra dozen cups and saucers.

"She loved to set a beautiful table and serve good food. She was a wonderful cook.

"**SHE LOVED** to read and had a library of 6,000 volumes--which will not be sold.

"The Kentucky farm would be opened to let people drive through on Sundays. On display below the swimming pool stood iron figures of Snow White and the Seven Dwarfs and Red Riding Hood and the Wolf."

Red Riding Hood and the Wolf will be auctioned.

ABOUT THE AUTHOR

Linda Light was born and reared in Waco, Texas. She attended the University of Texas at Austin, the Universities of Cologne and Hamburg in Germany on a Rotary International Scholarship for study abroad, and received her Master's Degree in Speech Communications and Acting from the University of Arkansas at Fayetteville. She has also attended Robert Redford's Sundance Film Institute outside of Salt Lake City, Utah, and the New York Film Academy.

Linda is a writer, actress, director, producer, and speaker, having worked in television, radio, and film. Ms. Light has written and performed her highly acclaimed one-woman show, **Pansy Yount of Spindletop**, since 1988 before standing ovation audiences throughout the country. Linda Light's best-selling first edition of *PASSIONS AND PREJUDICE: THE SECRETS OF SPINDLETOP* was on the "Top Ten Bestsellers List for Non-Fiction Books in the South" according to *The Oxford American Literary Magazine* for both 1995 and 1996. Producer Martin Jurow of the Academy Award Winning movie *Terms of Endearment* calls it "a good and great work, worthy of being made into a movie." This new, expanded special edition of *PASSIONS AND PREJUDICE: THE SECRETS OF SPINDLETOP,* published first in 1997, has been

PASSIONS AND PREJUDICE

released to commemorate the first time in history that the Second Spindletop Oil Gusher--the Yount Gusher--will be celebrated in Beaumont and Houston, Texas. In Beaumont, the celebration is scheduled on the date of the 72nd anniversary of the Second Spindletop Oil Gusher, November 13th, 1997. Miss Light's show **Pansy Yount of Spindletop** is scheduled to be presented on that evening at the Julie Rogers Theater in Beaumont, Texas. In Houston, Texas, Miss Light's show is scheduled to be a part of the Pin Oak Charity Horse Show in late October of 1997. Other dates are still to be announced at the time of this printing.

Miss Light has worked in two films, *Sylvester*, Martin Jurow of *Terms of Endearment*, producer, and *Bluegrass*, a made-for-T.V. mini-series, Alan Landsberg, Producer.

The actress/writer, now author, is currently working on making **Passions and Prejudice: The Secrets of Spindletop** into a movie as a vehicle to showcase and promote the American Saddlebred show horse beyond horse circles and before the public at large. She has written a new screenplay toward that purpose. Miss Light heads Spindletop Productions, Inc.

Linda is married to Dr. Theodor Langenbruch and has three children: Danielle, Andrea, and Christa. She has made her home in the beautiful and colorful Bluegrass area for the past seventeen years, and resides outside of Lexington in Richmond, Kentucky.

If you are interested in obtaining more information on the book, the performances, or the movie you may contact Miss Light at the following:

Spindletop Productions, Inc.
P.O. Box 806
Richmond, Kentucky 40476-0806

OR:
Phone: (606) 624-9726
Fax: (606) 623-1480
E-Mail: Spindletex@aol.com

For updates, we hope you will follow our Web site at **http://Spindletop2.com**.